DEFINITION OF THE ENGINEERING METHOD

By

Billy Vaughn Koen

American Society for Engineering Education
Washington, D.C.

Library of Congress Catalog Card No. 85-71927
ISBN 0-87823-101-3

Contents

Nevertheless, like a man who walks alone in the darkness, I resolved to go so slowly and circumspectly that if I did not get ahead very rapidly I was at least safe from falling.

—René Descartes
in *Discourse on the Method*

INTRODUCTION

The study of the engineering method is important to understand the world we have. The environment of man is a collage of engineering problem solutions. Political alliances and economic structures have changed dramatically as a result of the telephone, the computer, the atomic bomb and space exploration—all undeniably products of the engineering method. Look around the room in which you are now sitting. What do you find that was not developed, produced or delivered by the engineer? What could be more important than to understand the strategy for change whose results surround us now and, some think, threaten to suffocate, to pollute and to bomb us out of existence?

Yet, although we speak freely of technology, it is unlikely that we have the vaguest notion philosophically of what it is or what is befalling us as it soaks deeper into our lives. Were we asked, "What is the *scientific* method?" we would undoubtedly answer without difficulty. We might propose, "Science is theory corrected by experiment," or the other way around. With a bit more probing, we might explain the scientific method by developing Popper's theory of falsification or Kuhn's theory of paradigm shifts. But when asked, we, or anyone else for that matter, whether lay person, scientist or specialist in the history of science, would feel qualified to give a cogent response. Now, as we sit immersed in the products of the engineer's labor, we must ask: What is the *engineering* method?

The lack of a ready answer is not surprising. Unlike the extensive analysis of the scientific method, little significant research to date has sought the philosophical foundations of engineering. Library shelves groan under the weight of books by the most scholarly, most respected people of history analyzing the human activity called science.* No equivalent reading list treats the engineering method.

To identify a second reason for the lack of understanding of the engineering method, consider the professions that affect our daily lives, such as law, economics, medicine, politics, religion and science. For each we can easily name at least one person who is well-known to the general public as a wise, well-read scholar—a person to whom we can turn to put a profession in perspective. Now name an engineering statesman with similar qualifications. The challenge is to name an engineer who is wise, well-known, well-read and scholarly as an engineer. That is to say, in the event of a serious nuclear incident, the failure of the pylons on a large airplane, the pollution of ground water by chemical wastes or the failure of a walkway in a modern hotel, to whom can we or the news media turn to put the situation in perspective and settle our fears? The inevitable answer: To no one. No profession affecting the world to the extent engineering does can claim this isolation.

Unfortunately, the situation is far worse than just the lack of an engineering spokesman. Remembering that: 1) high school students do not take courses in engineering; 2) the study of technology is not required for a liberal arts degree; and 3) sociologists, psychologists, historians and religious proponents, not engineers, write most of the pro- and anti-technology literature—can we be sure, as the engineer speaks of optimization, factors of safety and feedback, that the lay person would understand an engineering spokesman if he did exist? Not only is there little research into the theory of engineering, no recognized spokesman and no general education requirements in the field, but engineers themselves are chronically averse to writing about their world. That people do not understand the engineering method and are a bit frightened by technology is not really too surprising.

---

*Among many, many others we find the work of the Ionian philosophers (Thales, Anaximander and Anaximenes), where many feel the germ of the scientific method was first planted in the 6th century B.C.; of Aristotle in the *Organum;* of Bacon in the *Novum Organum;* of Descartes in *Discours de la Méthode;* of Popper in *The Logic of Scientific Discovery;* and of Kuhn in the *Structure of Scientific Revolutions.*

This discussion seeks to redress this situation. It is in three parts, as follows:

Part I *Some Thoughts on Engineering*: This part describes the problem situation that calls for the talents of the engineer.

Part II *The Principal Rule of the Engineering Method:* Here the engineering method is defined.

Part III *Some Heuristics Used by the Engineering Method:* This section lists techniques engineers use to implement their method.

PART I
SOME THOUGHTS ON ENGINEERING

The use of the engineering method rather than the use of reason is mankind's most equitably divided endowment. By the *engineering method* I mean *the strategy for causing the best change in a poorly understood or uncertain situation within the available resources;* by *reason,* I mean the "ability to distinguish between the true and the false" or what Descartes has called "good sense." Whereas reason had to await early Greek philosophy for its development—and is even now denied in some cultures and in retreat in others—the underlying strategy that defines the engineering method has not changed since the birth of man.

The first objective of this chapter is to prepare the way for a consideration of the strategy the engineer uses to solve problems that will be given in Part II. Then attention shifts to the characteristics of a problem that requires the talents of this new acquaintance.

The Engineer

Most people think of engineers in terms of their artifacts instead of their art. As a result they see diversity where they should see unity. The question, "What is an engineer?" is usually answered by such a statement as "a person who makes chemicals, airplanes, bridges or roads." From the chemicals, the lay person infers the chemical engineer; from the airplanes, the aeronautical engineer; and from the bridges and roads, the civil engineer. Not only the lay person, but also the engineer makes this mistake. Because the pairing of engineers with their completed design is so enduring and the pairing with their use of method so fleeting, people insist they are engineers based on what they produce, regardless of how they go about it, instead of insisting they are engineers based on how they go about it, regardless of what they produce. But behind each chemical, each road, each pot, hides the common activity that brought it into being. It is to this unity of method we must look to legitimize the word *engineer.*

Characteristics of an Engineering Problem

Let us look in detail at the key words *change, resources, best* and *uncertainty* that have appeared in the definition of an engineering problem situation. Of these four, the reason for including the first two is relatively easy to explain. The next one is less well understood and must occupy more of our time. Throughout the discussion of the first three key words we will sense the fourth, the *lack of information* or *uncertainty* that always pervades an engineering problem, menacing in the wings. After the stage is set, the engineering strategy itself will make an appearance in Act II.

Change

Engineers cause *change.* The engineer wants to change, to modify or convert the world represented by one state into a world represented by a different one. The initial state might be San Francisco without the Golden Gate Bridge; the final state, San Francisco with this bridge. The initial state might be the Nile without a dam; the final one, the Nile with a new dam in place. Or the initial state might be a Neanderthal contemplating the death of a loved one; the final one, the world after the construction of a sepulcher. Graphically, each of these examples is represented in Figure 1, where time is given on the horizontal line or axis, and some measure of change in the world, on the vertical. The engineer is to cause the transition from *A* to *B.* To identify a situation requiring an engineer, seek first a situation calling for change.

We immediately run into three practical difficulties when we con-

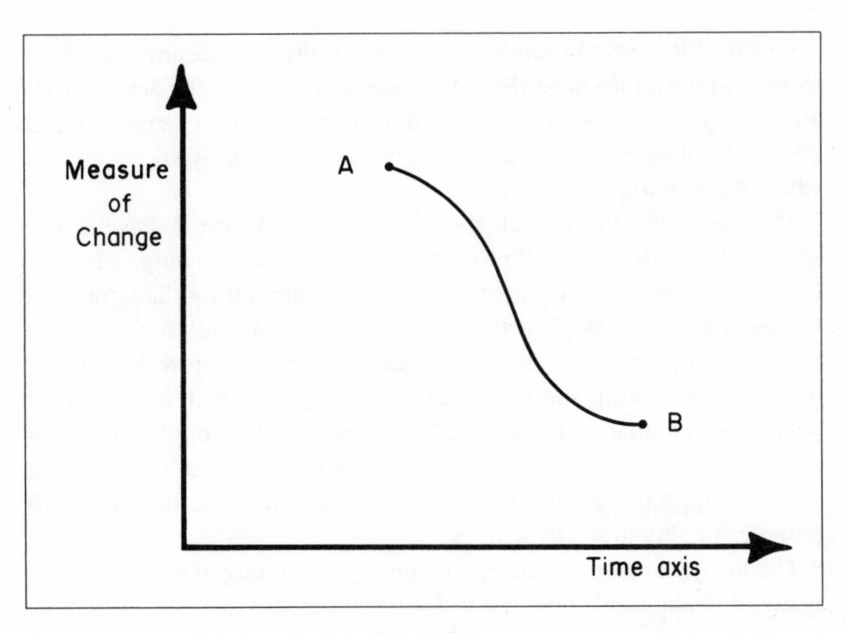

Figure 1

sider the engineer's change: the engineer doesn't know where he is going, how he is going to get there or if anyone will care when he does. Initially, the engineer is located at point *A* in Figure 1. The exact final state, point *B*, is not known at the beginning of the problem. An example will make this point clear. The Aswan High Dam in Egypt has increased the salinity of the Nile by 10 percent, has led to the collapse of the sardine industry in the Delta, has caused coastal erosion and has forced the 100,000 Nubians displaced by the reservoir to try to adapt to life as farmers on the newly created arable land. These liabilities have been balanced—some would say more than offset—by other assets, such as the generation of enough hydroelectric power to furnish half of Egypt's electrical needs. Our interest, however, is not to critique this spectacular engineering project or to reconcile conflicting opinions as to its net worth, but to emphasize that before construction, at state *A*, the engineer could not predict the exact change in salinity and erosion or the exact human costs to the sardine fishermen and the Nubians. The final state always has a reality the initial state lacks. Similarly, the order to "put a man on the moon by the end of the decade" lacks the specificity of the ladder Neil Armstrong descended to leave his footprint on the moon. The engineer is willing to develop a transition strategy but rarely is given a specific, well-defined problem to solve. Instead he must determine for himself what the actual problem is on the basis of society's diffuse desire for change. At the beginning of an engineering project, the engineer rarely knows exactly where he is going.

The next difficulty is with the change itself. Figure 1 falsifies the ease in deciding which path to take from *A* to *B* by showing only one. Usually a number of alternatives exist, each limited by different constraints. By definition, poverty could be easily eliminated in the United States by supplementing the income of each person below the poverty line. But dedicating the entire gross national product to this effort would be an unacceptable transition strategy. The engineer is not responsible for implementing a single given change, but for choosing the most appropriate one. In other words, at state *A* he doesn't know how he is going to get to state *B*.

The final difficulty in causing change is that an engineering goal has a way of changing throughout a design. From the start of a project to completion is often a long time. At present, for example, it takes 12 years to construct a nuclear reactor in America. During the completion of an engineering project, changes in the final goal often occur, requiring a reorientation of the project in midstream. In the automobile industry the public's demand has flitted from desire for a powerful automobile, to a safe automobile, to a small, fuel-efficient one—shifts so rapid that a new automobile design is often obsolete before it leaves the drawing board. With the lack of information about point *B* and the desired transition path between *A* and *B*, combined with changes in point *B* throughout the project, how can the engineer ever hope to cause the change he desires? Change is recognized as a characteristic of an engineering problem, but with all the attendant uncertainty, what strategy does the engineer use to achieve it?

Resources

The second characteristic of a situation that requires the services of an engineer is that the desired solution must be consistent with the *available resources.* Unfortunately, the engineer cannot select the best path from all conceivable transitions from the initial state to the final one. Physical, economic and political constraints always exist. (In spite of its favorable corrosion properties, no consideration was given to building the Golden Gate Bridge of an alloy of pure gold—for obvious reasons.) In the second place, then, the engineer always seeks the best change within the available resources.

These resources are an integral part of the problem statement and both define and constrain its solution. Different resources imply different problems. To make this point, one of my former professors would begin each class with a simple problem to be answered in fifty seconds by what an engineer would call a *back of the envelope* calculation. Once, for example, we were asked to estimate the number of ping-pong balls that would fit in the classroom. In addition to developing the ability to manipulate large numbers in our heads, these problems taught the importance of resources in the definition of a project. In fifty seconds we were to provide an answer—a correct engineering answer. Had we been given two days to respond, we would have been expected to measure the room and calculate the number—again, an entirely correct engineering response. I suppose if we had been given even more time, we could have filled the room with ping-pong balls and counted them. Though obviously similar, each of these problems was fundamentally different, as evidenced by its need for a different solution method. The answer to each was absolutely correct from an engineering point of view when both problem and time constraints were considered together.

Contrast an engineering problem to a scientific one with respect to each problem's dependence on resources. Although Newton was limited in the amount of time he had to develop his theory of gravitation and a modern cancer researcher is constrained by available funds, we usually think of each as trying to read the already-written book of nature instead of creating a new best-seller based on the available resources. We quibble, by extending the analogy beyond its bounds, if we assert that nature, and by implication, science, has a correct answer to the ping-pong ball problem and that the engineer is limited only by available resources in approximating this number. A similar sense of convergence to truth does not usually exist in an actual engineering problem. For example, if we try to argue that nature has an absolutely correct answer as to whether the Aswan High Dam should have been built and that the engineer will find it with additional resources, we quickly become inundated in profound philosophical water. Instead of looking for *the* answer to a problem, as does the scientist, the engineer seeks *an* answer to a problem consistent with the resources available to him. This distinction will become clearer when we now consider the engineer's notion of *best.*

The Best

The next characteristic of a problem situation requiring the engineer is that the solution should be the *best* or what is technically called the *optimum* solution. From an engineering point of view not all changes or all final states are equally desirable. Today few would suggest replacing the Golden Gate Bridge as a means of crossing San Francisco Bay with one of the wooden covered bridges that were once commonly seen in Maine. To identify a situation calling for the engineer, we must look for one in which not just any change, but the best change, is desired.

Best is an adjective applied redundantly to an existing engineering design. That a specific automobile exists proves that it is some engineer's subjective notion of the best solution to the problem he was given to solve. Saying that a Mercedes is a better automobile than a Mustang is incorrect if *better* is being used in an engineering sense. They are both optimum solutions to different specific design projects. Likewise, the complaint that "American engineers cannot build an automobile that will last for fifty years" can only be voiced by a person with little understanding of engineering. To construct such an automobile is well within the ability of modern automotive engineers, but to do so is a different design problem from the one currently given to the American engineer. It does make sense to prefer one design project over the other. An engineer could conceivably argue that designing an automobile similar to the Mercedes is a better goal than designing one similar to the Mustang, because it would last longer, conserve natural resources, promote national pride, or whatever. And, of course, a second engineer may feel that he could have produced a better final product than the first engineer given the same problem statement. But for the engineer who designed the Mustang, the automobile you see before you is his best solution to the problem he was given to solve. To exist is to be some engineer's notion of best.

Unlike science, engineering does not seek to model reality but society's perception of reality, including its myths and prejudices. If a nation feels that a funeral pyre should be aligned in a north-south direction to aid the dead person's journey to heaven, the model to be optimized will incorporate this consideration as a design criterion, regardless of the truth of the claim. Similarly, the engineering model is not based on an eternal or absolute value system, but on the one thought to represent a specific society. In a society of cannibals, the engineer will try to design the most efficient kettle. As a result, the optimum obtained from this model does not pretend to be the absolute best, but only the best relative to the society to which it applies. Contrast this with a scientific model. Speaking of Einstein's theory as the best available analysis of time and space implies that it comes closest to describing reality. It is better than the formulation of Newton because it explains more accurately or more simply our observations of nature. Best for the scientist implies congruence with an assumed external nature; best for the engineer implies congruence with a specific view of nature.

The appropriate view of nature for optimization is not just an objective, faithful model of society's view, but includes criteria known only to the engineer. One important consideration in lowering the cost of an automobile, for instance, is its *ease of manufacture.* If standard parts can be used or if the automobile can be constructed on an assembly line instead of by hand, the cost of each unit goes down. Ease of manufacture is a criterion seldom considered by the public but one that is essential to an accurate model for determining best automobile design. Because of these additional variables, the appropriate optimization model is not just a surrogate for society but includes subjective considerations of the engineer who makes the design.

In general, the optimum shifts when an optimization space with a reduced number of criteria is used. The best automobile based on the *axis system* of the public and that of the engineer will therefore differ. The person who criticized the engineer for not providing an automobile to last fifty years was making the error of not using a complete axis system. He was almost certainly not considering the ease with which such an automobile could be manufactured. As mentioned before, the design of a long-lasting automobile is possible and would be as exciting a challenge to the engineer as the present line of products. But the demand for it in the United States is so low, and the cost of producing it so high, that the cost per vehicle would be prohibitive. As a second example of a deficient system of axes being used by some members of the public, consider the complaint communication engineers occasionally hear: "This holiday season all the phone lines were busy and I couldn't get through. You would think the people at the phone company could anticipate the rush." Again we have two different axis systems being used—one by the lay person, one by the engineer.

The engineer could easily design a telephone system for the busiest period of the year, but the extra equipment that would be needed would remain idle the rest of the year and would have to be stored and maintained. The engineer uses an axis along which the cost of the extra, seldom-used equipment is traded off against the loss of service. The public has a right (I would say, obligation) to help select the problems for solution, the major design criteria, the return functions and the relative weights, but since the optimum shifts when an optimization space with a reduced number of criteria is used, it is naïve to criticize an engineer's optimum solution based on a reduced set of criteria without justifying the reduction.

Theoretically, then, *best* for an engineer is the result of manipulating a model of society's perceived reality, including additional subjective considerations known only to the engineer constructing the model. In essence, the engineer creates what he thinks an informed society should want based on his knowledge of what an uninformed society thinks it wants.

In some completed engineering projects we have experimental evidence that the axis system ultimately chosen as representative of society was deficient. The San Francisco Embarcadero freeway has become a classic example of the practical problem of trying to evaluate society's optimum. It was designed as the best way to move traffic about the city, money was appropriated and construction begun. The Embarcadero, now known as the freeway to nowhere, was abandoned in mid-construction because the design failed to include considerations that ultimately proved important. Criteria such as, "Don't block my view of the bay," "Don't raise the noise level or density of people in my neighborhood," and "Don't lower the overall quality of life," were important to the citizens of San Francisco. Too expensive to tear down, the Embarcadero now stands as a monument to the difference between engineering theory and engineering practice. Although theoretically the best design is determined once the optimization space is known, practically it is hard to be sure that we have not neglected an important axis in constructing this space. In the example mentioned, the optimization space of the engineer proved in practice a poor representation of society.

A fundamental characteristic of an engineering solution is that it is the best available from the point of view of a specific engineer. If this engineer knew the absolute good, he would do that good. Failing that, he calculates his best based on his subjective estimate of an informed society's perception of the good. With doubt about the criteria that are important to society, with doubt about the relative importance of these criteria and with doubt as to whether society's best reflects the individual's best, how can the engineer design the optimum product? What strategy does he use?

Best, change, uncertainty and *resources*—although we do not as yet know what the engineer's strategy is, it should not be too difficult to recognize a situation calling for its use. Unfortunately, it is. When the President of the United States promotes a new generation of space weapons to create a defensive umbrella and then calls on the "*scientific* community" to give us a way of developing it, he is confusing science and engineering. Relatively speaking, little new science is involved. Newton's Law of Gravitation, the equations of motion and the theory of energy emission by lasers or particle beams are all reasonably well understood by the scientist. If such a device is to be developed, the President would be better advised to call on the *engineering* community. Journalists share this confusion about what constitutes a scientific problem and what an engineering one. When reporters seeking information about the above-mentioned project went to "scientific experts" to evaluate the "feasibility of this space-age missile defense system," they went to the wrong place. Its feasibility is certainly more in doubt because of the difficulty of finding materials able to survive the tensile stresses, radiation damage and alien temperatures than because of something that violates the known laws of nature. If feasibility is the question, journalists should contact the dean of an engineering college, not their resident scientist.

Since confusion evidently exists in the mind of the non-engineer as to what constitutes an engineering problem, let us consider several examples with the defining characteristics of one in mind. The statement of an engineering problem might well be:

> I believe that this nation should commit itself to achieving the goal, before this decade is out, of landing a man on the moon and returning him safely to the earth. No single space project in this period will be more impressive to mankind, or more important for the long-range exploration of space; and none so difficult or expensive to accomplish. . . . [The cost would be] $531 million in 1962 and an estimated $7-9 billion over the next five years.

President Kennedy's statement fired the gun that sounded the start of one of America's most spectacular engineering adventures.

Engineering change is not limited to the creation of physical devices such as spaceships or highways. Other political, economic and psychological examples that require the engineer are easy to find. Some are almost trivial, others more complex—but all have the characteristics of an engineering problem. Perhaps a politician wants to be reelected or to win support in Congress for the construction of a dam in his home district; perhaps an economist would like to increase the gross national product or find a way to reduce the national debt; perhaps a psychologist would like to stop children from biting their nails or condition a race to create a utopian state using "behavioral engineering." The changes implied by these examples are usually not associated with the engineer, but careful study of the characteristics they share with the obvious engineering projects of designing a nylon plant, constructing a bridge across the Mississippi and building an electrical power station for New York City shows a definite pattern. For each, an engineer is needed.

If you, as with all humans since the birth of man, desire change; if the system you want to change is complex and poorly understood; if the change you will accept must be the best available; and if it is constrained by limited resources, then you are in the presence of an engineering problem. If you cause this change using the strategy to be given in the following pages, then you are an engineer.

PART II
THE PRINCIPAL RULE OF THE ENGINEERING METHOD

Chess is a complicated game. Although in theory a complete game tree can be constructed by exhaustive enumeration of all the legal first moves for the white side, then all possible responses to each by black, then white again and so forth until every possible game appears on the tree, in practice this procedure is impossible because of the enormous number of different moves and the limited resources of even the largest computer. Chess, therefore, defies analytical analysis.

To learn chess, a different strategy is usually needed to cause desirable change in our poor understanding of the game consistent with the available resources. This strategy consists of giving suggestions, hints and rules of thumb for sound play. For example:

1) Open with a center pawn,
2) Move a piece only once in the opening,
3) Develop the pieces quickly,
4) Castle on the king's side as soon as possible, and
5) Develop the queen late.

As we get better, we begin to hear:

6) Control the center,
7) Establish outposts for the knights,
8) Keep bishops on open diagonals, and
9) Increase your mobility.

Although these hints do not guarantee that we will win, although they often offer conflicting advice, although they depend on context and

Although these hints do not guarantee that we will win, although they often offer conflicting advice, although they depend on context and change in time, they are obviously better than trying to construct the game tree, and we do learn to play better chess. What is the name of this strategy?

One of the topics studied in a course in artificial intelligence is an unusual way of programming a computer to solve problems. Instead of giving it a program with a fixed sequence of deterministic steps to follow—an algorithm, as it is called—the computer is given a list of random suggestions, hints or rules of thumb to use in seeking the solution to a problem. The hints are called *heuristics*; the use of these heuristics, heuristic programming. Surprisingly, this vague, non-analytic technique works. It has been used in computer codes that play championship checkers, identify hurricane cloud formations and control nuclear reactors. Like the computer, both the method for solving its problem (learning to play chess) and that of the engineer in solving his problems (building bridges and so forth) depend on the same strategy for causing change. This common strategy is the *use of heuristics.* In the case of the engineer, it is given the name *engineering design.*

To analyze the important relationship between engineering design and the heuristic, four major objectives are set for this part of our discussion. They are to understand the technical term *heuristic*; to develop the engineer's strategy for change; to define a second technical term, *state of the art;* and finally, to state the principal rule for implementing the engineering method. The heuristic will be considered by definition, by examining its characteristics or signatures, and through its synonyms. Specific examples of engineering heuristics will be considered in Part III. The state-of-the-art will be explained by definition and by looking at its evolution and transmission from one generation of engineers to the next. Five examples proving the usefulness of this important engineering concept will be reviewed in this section.

The Heuristic

A Definition

A heuristic is anything that provides a plausible aid or direction in the solution of a problem but is in the final analysis unjustified, incapable of justification, and fallible. It is used to guide, to discover and to reveal. This statement is a first approximation, or as the engineer would say, a first cut, at defining the heuristic.

Signatures of the Heuristic

Although difficult to define, a heuristic has four signatures that make it easy to recognize:

- A heuristic does not guarantee a solution;
- It may contradict other heuristics;
- It reduces the search time in solving a problem; and
- Its acceptance depends on the immediate context instead of on an absolute standard.

Let us compare the presumably known concept of a scientific law with the less-well-known concept of the heuristic with respect to these four

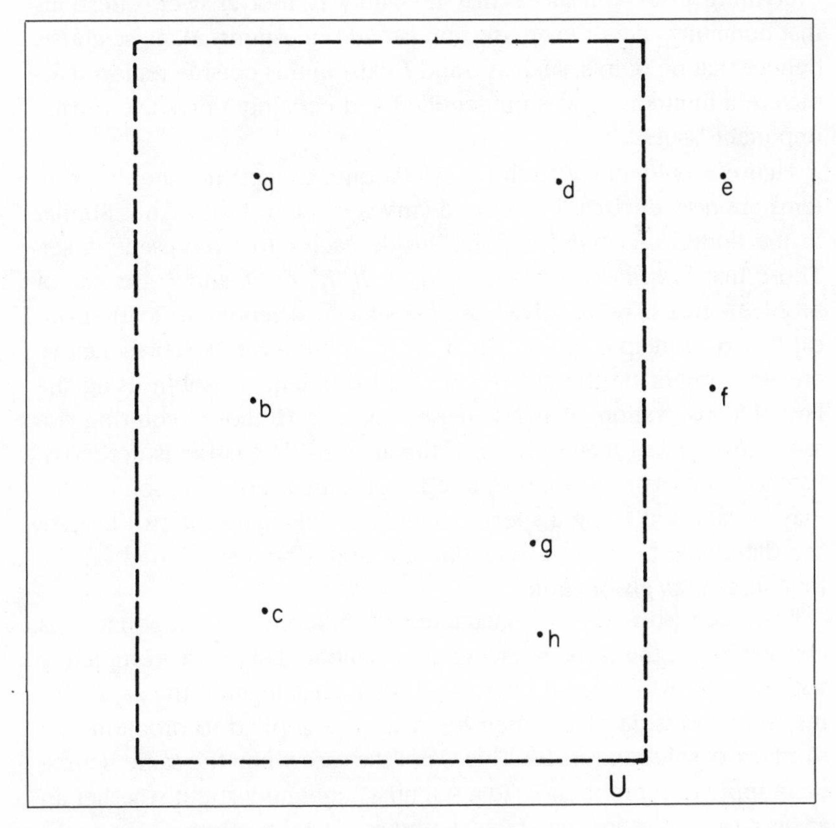

Figure 2

signatures. In doing so we will come to appreciate the rationality of using irrational methods to solve problems.

This comparison may be easier if we use the simple mathematical concept of a set. The interior of Figure 2 represents the set of all problems that can ultimately be solved. It will be given the name *U*. *U* is not limited to those problems solvable on the basis of present knowledge, but includes all problems that would theoretically be solvable given perfect knowledge and an infinite amount of time. The points labeled *a, b, c, d, g* and *h* are elements of *U* and represent some of these problems. If you prefer, it is sufficient for our present purposes to think of *U* as a simple list of questions about nature that humanity will someday be able to answer. On this list most people born into the Western tradition would include: Will the sun rise tomorrow? Does bread nourish? If I release this ball, will it fall? And, should the Aswan High Dam have been built? Outside this area is everything else—questions that humanity cannot answer, questions that humanity cannot even ask, and pseudo questions. Many scientists believe that no points, such as *e* and *f*, exist in this outside region. This picture admittedly leaves unidentified and certainly unresolved many important issues.

Figure 3 is identical to the previous one, except the sample problems are now encircled by closed curves labeled *A* through *I*. Similar to the dotted rectangle, the area inside each curve represents a set. Those that have been crosshatched, *A, B, B′, C, D,* and *I*, are sets of problems that may be solved using a specific scientific or mathematical theory, principle or law. Set *A*, with problem *a* as a representative element, might be the collection of all problems solvable using the law of conservation of mass-energy, and set *B*, those requiring the associative law of mathematics. If the area inside a curve is not crosshatched, such as *E, F, G* and *H*, it represents the set of all problems that may be attacked using a specific heuristic. This figure helps illustrate the difference between a scientific law and a heuristic based on the four signatures given earlier

First, heuristics do not guarantee a solution. To symbolize this characteristic, the sets referring to scientific laws rest completely within set *U*, while those referring to heuristics include the area both inside and outside of *U*. When heuristic *E* is applied to problem *d*, a satisfactory solution results. This is not the case when the same heuristic is applied to problem *e*. To a scientist, ambiguity about whether an answer to a question has been found is a fatal weakness. He seeks procedures, strategies and algorithms that give predictable results known to be true. Uncertainty about a solution's validity is a sure mark of the use of a heuristic.

Unlike scientific theories, two heuristics may contradict or give different answers to the same question and still be useful. This blatant disregard for the classical law of contradiction is the second sure signature of the heuristic. In Figure 3 the overlap of the two scientific sets, *C* and *D*, indicates that a problem in the common area such as *c* would require two theories for its solution. The need for both the law of gravitation and the law of light propagation to predict an eclipse is a good illustration. Since combinations of two, three and often more

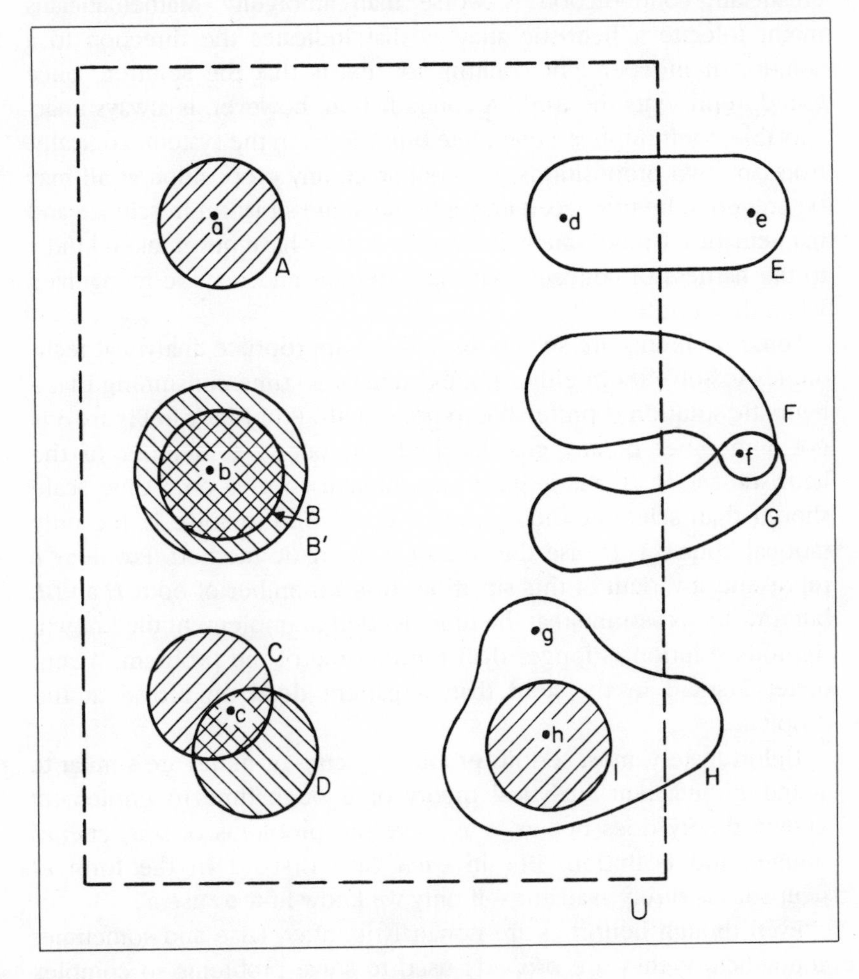

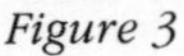
Figure 3

scientific and mathematical theories must work together to solve most problems, *U* is overlaid with a complex array of scientific sets.

This is not true in the case of the heuristic. Here, the overlap of *F* and *G* represents the conflicting answer given to problem *f* found outside of *U*. Although at times two heuristics might be needed to arrive at an answer and, hence, to overlap within *U*, the most significant characteristic of a heuristic is its rugged individualism and tendency to clash with its neighbors. We have already seen, for example, that at least three different heuristic strategies are available to arrive at the number of ping-pong balls in a room and that each leads to a different, but completely acceptable, engineering answer. For a mathematician, contradiction is worse than ambiguity. Mathematicians might tolerate a heuristic strategy that indicates the direction to a solution if independent confirmation exists that the solution, once found, represents the truth. A contradiction, however, is always unacceptable, for it implies a complete breakdown in the system. Logically, from any two propositions that contradict, any proposition at all may be proved to be true—certainly a bothersome situation in science and mathematics. Unlike scientific laws, heuristics have never taken kindly to the harness of conventional logic systems and may be recognized when they bridle.

Some problems are so serious and the appropriate analytical techniques to solve them either nonexistent or so time-consuming that a heuristic solution is preferable to none at all. Problem *g* in Figure 3 is not a member of any crosshatched set, but is a member of the heuristic set *H*. If *g* is lethal to the human species on a time scale shorter than scientific theory can be developed to solve it, the only rational course is to use the irrational heuristic method. Problem *h* represents a variant of this situation. It is a member of both *H* and *I*, but now let us assume that the time needed to implement the known, rigorous solution is longer than the lifetime of the problem. Again, better first-aid in the field than a patient dead on arrival at the hospital.

Unfortunately, most serious problems facing mankind are similar to *g* and *h*. Sufficient analytical theory or enough time to implement known theory does not exist to solve the problems of war, energy, hunger and pollution. But in each case first-aid in the form of heuristics is surely available—if only we know how to use it.

Even though heuristics are nonanalytic, often false and sometimes contradictory, they are properly used to solve problems so complex and poorly understood that conventional analytical techniques would be either inadequate or too time-consuming. This ability to solve unsolvable problems or to reduce the search time for a satisfactory solution is the third characteristic by which a heuristic may be recognized.

The final signature of a heuristic is that its acceptance or validity is based on the pragmatic standard—it works or is useful in a specific context—instead of on the scientific standard—it is true or consistent with an assumed, absolute reality. For a scientific law the context or standard of acceptance remains valid, but the law itself may change or become obsolete; for a heuristic the contexts or standards of acceptance may change or become obsolete, but the heuristic itself remains valid. Figure 3 helps make this distinction.

Science is based on conflict, criticism or critical thought, on what has been called the Greek way of thinking. A new scientific theory, say B', replaces an old one, B, after a series of confrontations in which it is able to show that—as an approximation to reality—it is either broader in scope or simpler in form. If two scientific theories, B and B', predict different answers to a question posed by nature, at least one of them must be wrong. In every scientific conflict there must be a winner. The victor is declared the best representative of "the way things really are" and the vanquished discarded as an interesting, but no longer valid, scientific relic. Ironically, the loser is often demoted to the rank of a heuristic and still used in cases of expediency. Thus, Einstein's theory replaced Newton's as scientific dogma, and Newton's Law of Gravitation is now used, in the jargon of the engineer, when a *quick and dirty* answer is needed. The scientist assumes that the set, U, exists, that it does not change in time, that it is eternal. Only the set of currently accepted scientific laws changes in time.

On the other hand, the absolute value of a heuristic is not established by conflict but depends exclusively on its usefulness in a specific context. If this context changes, the heuristic may become uninteresting and disappear from view, awaiting, perhaps, an eventual change of fortune. Unlike a scientific theory, a heuristic never dies; it just fades from use. A different interpretation of Figure 3 is therefore more appropriate in the case of a heuristic.

For the engineer the set U represents all problems he wants the answer to at a given moment instead of all problems that are ultimately answerable. As a result, it is not a constant but varies in time. The engineer's set U ebbs as the obsolescence of the buggy has left the heuristics for buggy whip design high and dry on the shelf in the blacksmith's workshop, and it flows as renewed interest in self-sufficiency has sent young people in search of the wisdom of the pioneers. One heuristic does not replace another by confrontation but by doing a better job in a given context. Both the engineer and Michelangelo "criticize by creation, not by finding fault."

The dependency on immediate context instead of absolute truth as a standard of validity is the final hallmark of a heuristic. It and the other three signatures are not the only important distinctions between the scientific law and the heuristic, but they are sufficient, I think, to indicate a clear difference between the two.

Synonyms for the Heuristic

Most engineers have never consciously thought of the formal concept of the heuristic, but all engineers recognize the need for a word to fit the four characteristics just given. They frequently use the synonyms rule of thumb, intuition, technique, hint, rule of craft, engineering judgment, working basis, or, if in France, *le pif* (the nose) to describe this plausible, if fallible, basis of the engineer's strategy for solving problems. Each of these terms captures the feeling of doubt characteristic of the heuristic.

This completes consideration of the technical word *heuristic* needed for a definition of the engineering method until Part III, where an extensive list of examples will be given. We have analyzed this important concept by analogy with the hints and suggestions given to learn chess, by definition, by looking at four signatures that distinguish it from a scientific law and by reviewing a list of its synonyms.

I hasten to add that neither the word *heuristic* nor its application to solving particularly intractable problems is original with me. Some historians attribute the earliest mention of the concept to Socrates about 469 B.C., and others identify it with the mathematician, Pappus, around 300 A.D. Principal among its later adherents have been Descartes, Leibnitz, Bolzano, Mach, Hadamard, Wertheimer, James, and Koehler. In more recent times, Polya has been responsible for its continued development.* Without a doubt, the study of the heuristic is very old. But as old as it is, the use of heuristics to solve difficult

* Polya, G., *How to Solve It,* Princeton University Press, 1945, 1973.

problems is older still. Heuristic methods were used to guide, to discover and to reveal a plausible direction for the construction of dams, bridges and irrigation canals long before the birth of Socrates.

Definition of the Engineering Method

What is original in our discussion is the definition of the engineering method as the use of engineering heuristics to cause the best change in a poorly understood situation within the available resources. This definition is not meant to imply that the engineer just uses heuristics from time to time to aid in his work, as might be said of the mathematician. Instead my thesis is that the *engineering strategy for causing desirable change in an unknown situation within the available resources* and the *use of heuristics* is an absolute identity. In other words, everything the engineer does in his role as engineer is under the control of a heuristic. Engineering has no hint of the absolute, the deterministic, the guaranteed, the true. Instead it fairly reeks of the uncertain, the provisional and the doubtful. The engineer instinctively recognizes this and calls his ad hoc method "doing the best you can with what you've got," "finding a seat-of-the-pants solution," or just "muddling through."*

State-of-the-Art

Instead of a single heuristic used in isolation, a group of heuristics is usually required to solve most engineering design problems. This introduces the second important technical term *state-of-the-art.* Anyone in the presence of an engineer for any length of time will have heard him slip this term into the conversation. He will proudly announce that his stereo has a state-of-the-art speaker system or that the state-of-the-art of computer design in his home country is more advanced than elsewhere. Since this concept is fundamental to the art of engineering, attention now shifts to the definition, evolution and transmission of the state-of-the-art, along with examples of its use.

* This definition of the engineering method was first presented in a paper entitled, "The Teaching of the Methodology of Engineering to Large Groups of Non-Engineering Students," Gulf-Southwest Section, American Society for Engineering Education, March 26, 1971.

A Definition

State-of-the-art, as a noun or an adjective, always refers to a set of heuristics. Since many different sets of heuristics are possible, many different states-of-the-art exist, and to avoid confusion each should carry a label to indicate which one is under discussion. Each set, like milk in the grocery store, should also be dated with a time stamp to indicate when it is safe for use. Too often, the neglect of the label with its time stamp has caused mischief. With these two exceptions, no restrictions apply to a set of heuristics for it to qualify as state-of-the-art.

In the simplest, but less familiar, sense, state-of-the-art is used as a noun referring to the set of heuristics used by a specific engineer to solve a specific problem at a specific time. The implicit label reminds us of the engineer and the problem, and the time stamp tells us when the design was made. For example, if an engineer wants to design a bookcase for an American student, he calls on the rules of thumb for the size and weight of the typical American textbook, on engineering experience for the choice of construction materials and their physical properties, and on standard anatomical assumptions about how high the average American student can reach and so forth. The state-of-the-art used by this engineer to solve this problem at this moment is the set of these heuristics. If the same engineer were asked to design a bookcase for a French student, he would use a different group of heuristics and hence a different state-of-the-art. (Bookcase design is not the same in the United States and France.) Now consider two engineers who have been given the same problem of designing a bookcase for an American student. Each will produce similar, but different, designs. Since a product is necessarily consistent with the specific set of heuristics used to produce it, and since no two engineers have exactly the same education and past experience, each will have access to similar, but distinctly different, sets of heuristics and hence will create a different solution to the same problem. State-of-the-art as a noun refers to the actual set of heuristics used by each of these engineers.

In a complicated, but more conventional, sense, state-of-the-art also refers to the set of heuristics judged to represent "best engineering practice." When a person says that his stereo has a state-of-the-art speaker system or that he has a state-of-the-art bookcase, he does not just mean that they are consistent with the heuristics used in their design. That much he takes for granted. Instead he is expressing the stronger view that a representative panel of qualified experts would judge his speaker system or his bookcase to be consistent with the best set of heuristics available. Once again, state-of-the-art refers to an identifiable set of heuristics.

Because the design of a bookcase requires only simple, unrelated heuristics, it misrepresents the complexity of the engineering state-of-the-art needed to solve an actual problem. More typically, the state-of-the-art is an interrelated network of heuristics that control, inhibit and reinforce each other. For example, the physical property of a large organic molecule called the enthalpy may be determined either in the laboratory (one heuristic) or by estimating the number of carbon and hydrogen atoms the macromolecule contains and then applying a known formula (a rival heuristic). In practice, the engineer uses sometimes one method, sometimes the other. Obviously he has another heuristic—perhaps something like "go to the laboratory if you need 10 percent accuracy and have $5,000"—to guide his selection between the two. Bookcase design and even the determination of the enthalpy of a large organic molecule are such simple examples that they hardly suggest the complexity of a state-of-the-art. It is to your imagination I must finally turn to visualize how much more complicated the state-of-the-art used in the design of an airplane must be as the heuristics of heat transfer, economics, strength of materials and so forth influence, control and modify each other. Whether it is the set that was actually used in a specific design problem or the set that someone feels would be the best, state-of-the-art always refers to a collection of heuristics—most often a very complicated collection of heuristics at that. Its imaginary label must let us know which engineer, which problem, and which set are under consideration.

The state-of-the-art is a function of time. It changes as new heuristics become useful and are added to it and as old ones become obsolete and are deleted. The bookcase designed for a Benedictine monk today is different from the one designed for St. Benedict in 530 A.D. When we discussed the design of a bookcase for students, I did not emphasize the time stamp that must be associated with every state-of-the-art. Now is the time to correct this omission and consider the evolution of a set of heuristics in detail, beginning with a well-documented example.

Evolution

In overall outline, scholars feel that the evolution of the present-day cart took place in a series of stages. Since the wheel probably evolved from the roller, it is assumed that the earliest carts had wheels rigidly mounted to their axles, wheels and axle rotating as a unit. This assembly has the disadvantage that one wheel must skid in going around a corner. As an improvement, the second stage was probably a cart with the axle permanently fixed to the body and with each wheel rotating independently. In this design neither the wheels nor the axle were capable of pivoting. It has the disadvantage that the cart cannot go around corners unless forced into each new position. This same difficulty still bothers parents who own the old-fashioned baby stroller with fixed wheels.

After 20 or 30 centuries, the engineer learned how to correct this problem by allowing the front axle to pivot on a king bolt as stage three in the evolution of cart design. Since the front and back wheels were large and the same size, this cart could not turn sharply without the front wheels scraping against its body. In the final stage, the front wheels were reduced in size and allowed to pass under the bed of the wagon as the front axle pivoted. This process of evolution has continued into the present day, of course, as the cart has become the automobile. But as we are short on time and this modern state-of-the-art is more complicated than you might think, you will have to ask your local racing enthusiast or mechanical engineer what rack-and-pinion steering is and how it works.

Transmission

Down through the ages the state-of-the-art has been preserved, modified and transmitted from one individual to another in a variety of ways. The earliest method was surely a simple apprentice system in which artisans carefully taught rules of thumb for firing clay and chipping flint to their assistants who would someday replace them. With hieroglyphics, cave paintings and, later, books, the process became more efficient and was no longer dependent on a direct link between teacher and taught. Finally, in more recent times, trade schools and colleges began to specialize in teaching engineers. In spite of the importance of apprentice, book and school in preserving, modifying and transmitting accumulated engineering knowledge, they need not detain us longer because of their familiarity.

An additional method is less well-known and worth a minute's delay. If an engineer were asked to design a cart today, he would use an articulated front axle and wheels that were small enough to pass under the bed of the cart—not because he is familiar with the evolution of cart design, but because the carts he actually sees around him today are constructed this way. All traces of the state-of-the-art that dictated axle and wheels turning as a unit have disappeared. The engineer does not need to know the history of cart design; the cart itself preserves a large portion of the state-of-the-art that was used in its construction. In other words, modern design does not recapitulate the history of ancient design.

Since the heuristics of tomorrow are embodied in the concrete objects of today, the engineer is unusually sensitive to the physical world around him and uses this knowledge in his design. I once asked an architectural engineer to estimate the size of the room in which we had been sitting during an earlier meeting. He quickly gave an answer by remembering that the room had three concrete columns along the side wall and two along the front. Since he also knew the rule of thumb for standard column spacing that applied to a room such as we had been in, he could calculate its size quite accurately. This awareness of the present world translates directly into the heuristics used to create a new one. If a proposed room, airplane, reactor or bridge deviates too far from what he has come to expect, the engineer will question, recalculate and challenge. Although he may be unaware that he is wearing glasses, the engineer judges, creates and sees his world through lenses ground to the prescription of his state-of-the-art.

To review: the state-of-the-art is a specific set of heuristics designated by a label and time stamp. It changes in time and is passed from engineer to engineer either directly, by the technical literature, or in a completed design. Typically it includes heuristics that aid directly in design, those that guide the use of other heuristics, and, as we will see later, those that determine an engineer's attitude or behavior in solving problems. The state-of-the-art is the context, tradition or environment (in its broadest sense) in which a heuristic exists and based upon which a specific heuristic is selected for use. We might also characterize the state-of-the-art of the engineer as his privileged point of view.

State-of-the-art, no matter how it is written, is both a cumbersome and inelegant term. After only a few pages of definition, we have become tired of seeing it in print. Therefore, from now on I propose to replace it by its acronym *sota*. Coining a new word by using the first letters in an expression's constituent words is a familiar procedure to the engineer, who speaks freely of radar (*ra*dio *d*etecting *a*nd *r*anging) and the lem (*l*unar *e*xcursion *m*odule). In large engineering projects such as the manned landing on the moon, acronyms are so frequently used that often a project description appears to the non-engineer as written in a foreign language. At times an acronym even takes on a life of its own, and we forget the words used to create it. Few remember the original definitions of laser, scuba, zip code and snafu, and I wish the same fate for this new acronym *sota*. From now on, sota, used both as adjective and noun, is to be taken as a technical term meaning an identifiable set of heuristics.

Example Uses of an Engineering Sota

Without due consideration, the concept of an engineering state-of-the-art as a collection of heuristics appears contrived, and its acronym, gimmicky. The frequency with which the word *sota* will appear in the remainder of this discussion and the relief we will feel at not seeing its expanded form each time will answer the second criticism. Five specific examples showing the effectiveness of the sota as a tool for bringing understanding to important aspects of the engineering world will answer the first. Various sets of heuristics will now be used to:

1) Compare individual engineers,
2) Establish a rule for judging the performance of an engineer,
3) Compare the technological development in various nations,
4) Analyze several pedagogical strategies of engineering education, and
5) Define the relationship between the engineer and society.

This last example will also suggest the importance of technological literacy for the non-engineer and liberal literacy for the engineer. Although these examples will occupy a reasonable amount of our time, they should dispel any feeling of artificiality in the notion of a sota and suggest important areas for future research. We will then, at last, be in a position to consider the principal rule for implementing the engineering method.

COMPARISON OF ENGINEERS:

The individual engineer, in his role as engineer, is defined by the set of heuristics he uses in his work. When his sota changes, so does his proficiency as an engineer. This set will be represented by the area

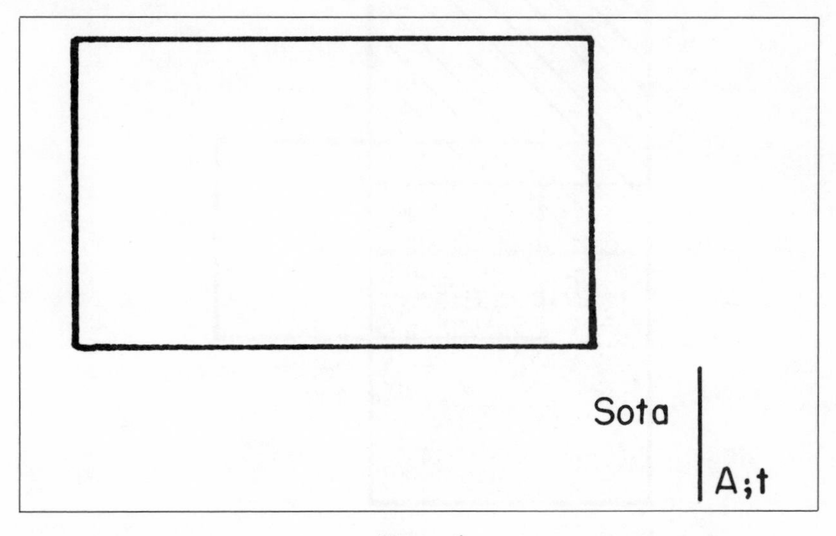

Figure 4

inside of the closed curve in Figure 4. Characteristic of all sotas, this one needs a label to differentiate it from others and a time stamp to indicate when it was evaluated. Using the symbol, sota $|_{A;t}$, for the state-of-the-art of Mr. *A* at time *t* will greatly simplify the discussion. The symbol representing my current definition as an engineer is, of course, sota $|_{Koen; 1985}$.

No two engineers are alike. The first example we will consider of the use of a sota will make this point. The sotas of three engineers, *A*, *B*, and *N*, are shown in Figure 5. They share those heuristics inside the area indicated by the small rectangle where they overlap, but each also encloses additional area to account for the unique background and experience of the engineer it represents. In general, if *A*, *B*, and *N* are all civil engineers, the overlap of their sotas is larger than if *A* is a civil engineer, *B* is a chemical engineer, and *N* is a mechanical engineer. Most civil engineers have read the same journals, attended the same conferences and quite possibly used the same textbooks in school. Not surprisingly, they share many engineering heuristics.

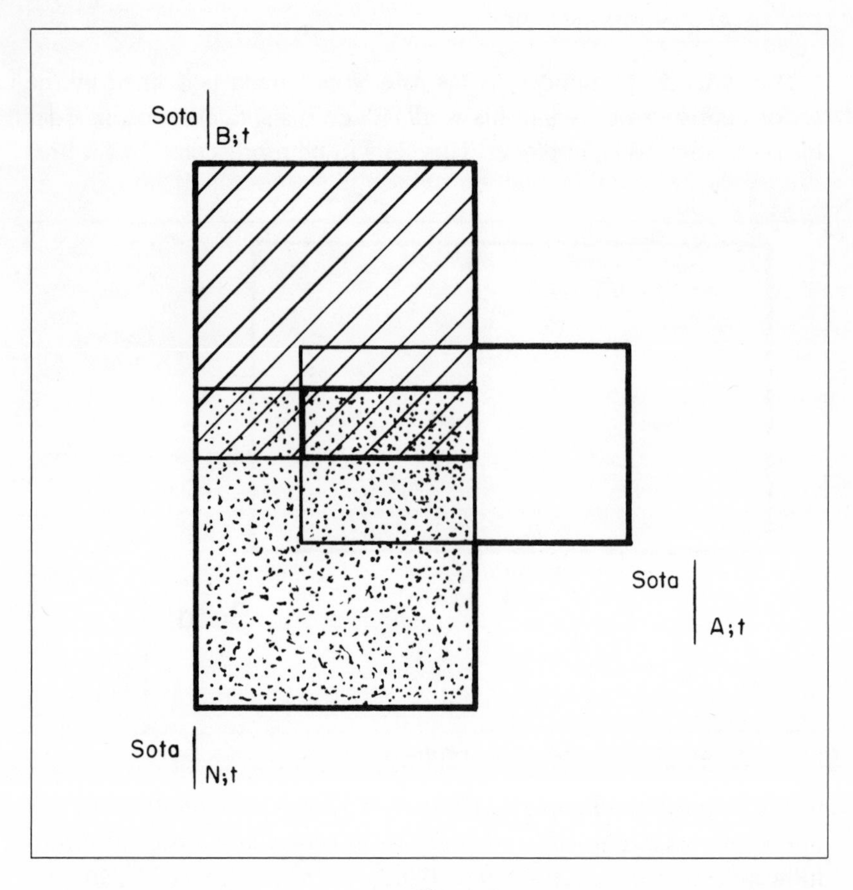

Figure 5

If the sotas of all modern civil engineers, instead of only the three, were superimposed, the common overlap of all of these sets, technically called the intersection, would contain the heuristics required to define a person as a modern civil engineer. Both society and the engineering profession have a vested interest in preserving the integrity of this area. The label *civil engineer* must insure a standard approach and minimum level of competence in solving civil engineering problems. What is now needed is research to determine the minimum intersection necessary to certify an engineer as an expert in, say, heat transfer, hydraulics or strength of materials. We will return to the intersection of all engineering sotas later in our discussion to give the heuristics it must contain for a person to be properly called an called an engineer. For now, remember that the overlap of the appropriate sotas measures the similarity and dissimilarity of engineers. It is an instructive example of the use of the concept *sota.*

No one, I think, would argue that engineers should not be held responsible for their work. The difficulty is knowing when they have done a satisfactory job. A discussion of the correct rule for judging the engineer will be a good second example of using an engineering sota and will emphasize, once again, the importance of attaching an imaginary label and a time stamp to each one. This discussion will also point out the difficulty of implementing this Rule of Judgment.

RULE OF JUDGMENT:

The heavy outside border in Figure 5 indicates the set of all heuristics used by *A*, *B*, and *N*, or sota $|_{A,B,N;t}$. If all engineers were included in this figure, it would delimit the sota of the engineering profession as a whole, or sota $|_{eng.prof.;t}$. The stippled area in Figure 6 reproduces this sota at a given time, *t.*

No engineer will have access to all of the heuristics known to engineering, but in principle some engineer somewhere has access to each heuristic represented in this figure. The black solid circle represents the subset of heuristics needed to solve a specific problem. The combined wisdom of the engineering profession defines the best possible engineering solution. This overall sota represents *best engineering practice* and is the most reasonable practical standard against which to judge the individual engineer. It is a relative standard instead of an absolute one, and like all sotas it changes in time.

To my knowledge, no engineers are clairvoyant. Handicapped in this way, it would seem unreasonable to expect them to make a decision at one moment based on information that will only become available later. They can only make a decision based on the set of heuristics that bears the time stamp certifying its validity at the time the design must be made. With these considerations in mind, we can formulate the fundamental Rule of Judgment in engineering: Evaluate an engineer or an engineering design against the sota that defines best practice at the time the design was made.

This rule is logical, defensible and easy to state. Unfortunately, it is not universally applied owing to ignorance, inattention and a genuine difficulty in extracting the sota that represents good engineering

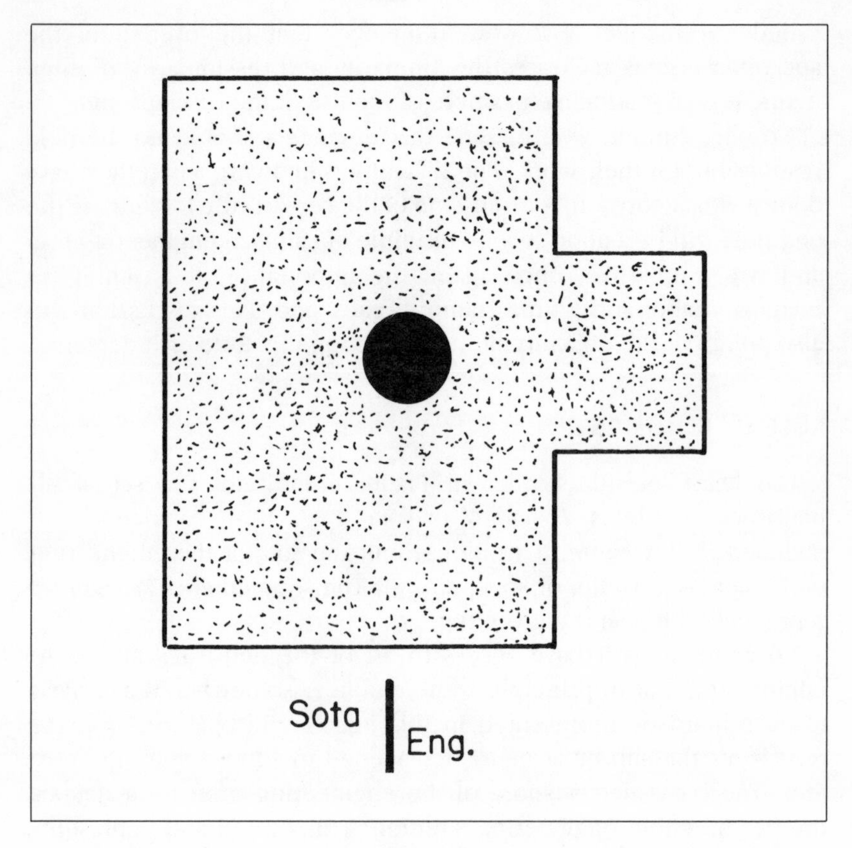

Figure 6

practice from the set of all engineering heuristics in a specific case. Each of these three reasons for not observing the Rule of Judgment is worthy of special attention.

For the lay person, the failure of an engineering product usually means that some engineer somewhere has done a poor job of design. This criticism is based on ignorance of the correct basis for judging the engineer and is indefensible for two reasons. First, an engineering design always incorporates a finite probability of failure. The engineer uses a complex network of heuristics to create the new in the area of uncertainty at the margin of solvable problems. Hence some failures are inevitable. Had ancient engineers remained huddled in the security of the certain, they would never have ventured forth to create the wheel or the bow. The engineer should not be criticized by looking only at a specific failure and ignoring the context or sota that represents the best engineering practice upon which decisions leading to that failure were based. Second, the world changes around the completed engineering product. A sota tractor on the date it was delivered is not necessarily appropriate for working steeper terraces and pulling heavier loads fifty years later, and it may fail or overturn or both. Often the correct basis for judging the performance of an engineer is not used because the public, including juries charged with deciding cases involving product liability and journalists reporting major technological failures, does not know what it is.

Engineers are also misjudged through inattention. Since the sota is a function of time, special attention is needed to ensure that the engineer is evaluated against the one valid at the time he made his design. Two examples will demonstrate how easy it is to forget this requirement, easy even for an engineer.

One of my French engineering colleagues, undoubtedly carried away with nationalistic zeal, surveyed the modern Charles de Gaulle Airport in Paris and explained that he had just had a miserable time getting through O'Hare Airport in Chicago. He then added that he could never understand why American engineers are not as good at designing airports as are the French. His mistake was wanting a clairvoyant engineer. Leaving aside the factor of scale—the American airport was at the time the busiest in the world; de Gaulle had been in operation only two weeks—two facts are beyond dispute: each airport was consistent with the sota at the time it was built and, given the intimate exchange of technical information at international technical meetings, the sota on which the French airport was based was surely a direct outgrowth of the earlier one used in Chicago. Even an engineer sometimes forgets the time stamp required on an engineering product.

The second example concerns the aphorism of the American frontier that a stream renews itself every ten miles. Essentially this means that a stream is a buffered ecosystem capable of neutralizing the effects of an incursion within a short distance. Let us assume that an enterprising pioneer built a paper mill on a stream and into it discharged his waste. According to the above rule of thumb, no damage was done. Now let us add that over subsequent decades additional mills were constructed until the buffering capacity of the stream was exceeded and the ecosystem collapsed. Your job is to fix blame. If you argue that later engineers were wrong to use a heuristic that was no longer valid, I might agree. Presumably the original heuristic is applicable only to a virgin stream. If you argue that the original plant should be modified to make it consistent with later practice, a process the engineer calls *retro-* or *back-fitting,* I might also agree, although I am curious whether you would require the pioneer, later plant owners or society to pay for this often expensive process. But if you criticize the original decision to build a plant on the basis of today's set of heuristics, I most certainly do not concur. As others have said, we must judge the past by its own rule book, not by ours.

These two examples show how easy it is to forget the factor of time in engineering design through inattention. But now let us consider a more troublesome reason why the engineer is often not judged on the basis of good engineering practice at the time of his design. The problem is agreeing on what sota is to be taken as representative of good engineering practice in a specific case. All engineers cannot be asked for their opinions; that is, sota $|_{\text{eng. prof.}}$ cannot be used as a standard. The only recourse is to rely on a "panel of qualified experts" to give its opinion. But how is such a panel to be constituted? Is membership to be based on age, reputation or experience? In determining the set of heuristics to represent a sota chemical plant, should foreign engineers be consulted? And finally, when best engineering practice is used as a basis for how-safe-is-safe-enough for a nuclear reactor, should members of environmentalist groups be included? No absolute answers can be given. But the engineer has never been put off by a lack of information and is willing to choose the needed experts—heuristically. Like any other sota, the set of heuristics he uses to choose his panel will vary in time and must represent good engineering practice at the time he constitutes it.

Agreement about the sota that is to represent best engineering design in the present is hard enough, but agreement about the set of heuristics appropriate fifty years ago is even harder. Many of the designers of engineering projects still in use are no longer living. Was the steel in the Eiffel Tower consistent with the best engineering practice of its day? With no official contemporary record to document good engineering judgment, history easily erases the engineering profession's memory as to what was the appropriate sota for use in the past. Given the recent rash of product liability claims against the engineer, what is now needed is an archival sota $|_{\text{best. eng. judg.}}$ to allow effective implementation of the Rule of Judgment.

RELATIVE TECHNOLOGICAL DEVELOPMENT OF COUNTRIES:

The sota $|_{eng.prof.;t}$ evaluated for a country is obviously a measure of its technological development and ability to solve technological problems. It will serve as a third example of the importance of the technical word *sota*. A country without access to the sota that represents best engineering practice is at a definite disadvantage. Underdeveloped countries are underdeveloped precisely for this reason. Even among technologically advanced countries subtle differences in sota result in significantly different products. Competent nuclear engineers have recently reported that a wide variation in testing philosophy in the design of the so-called fast nuclear reactor is evident among major nuclear powers. They report that Americans do extensive testing of design variations and actual components before building the reactor itself; that the French do extensive testing on the full-scale reactor (with the British doing significantly less); and that the Russians prefer to build the reactor first and then see if it can be made to work.

I will refer to the difference between the American and Russian philosophies later and give a name to this heuristic. For now it is sufficient to recognize that different countries use different sotas when it comes to the testing philosophy of fast nuclear reactors, and that testing philosophy inevitably affects the final product. A colleague once told me he was absolutely convinced that an American engineer was the first to step on the moon because the National Aeronautics and Space Administration required more attention to quality control of individual components than did its competitors. Whether this is true or only the exaggeration of yet another engineer carried away by nationalistic zeal, the country with the most effective heuristics is clearly the most advanced technologically and the best able to respond to new technological challenges. What is needed is research to determine if the sota that represents best engineering practice in America is consistent with the sota that represents best engineering practice worldwide.

As an addendum, I cannot help but wonder if someday American engineers who typically speak only English and base their designs only on the heuristics encoded in English, will not find themselves at a serious competitive disadvantage with respect to multilingual engineers who base their designs on a sota containing heuristics encoded in a variety of languages.

ENGINEERING EDUCATION:

Let us look at the problem in engineering education caused by the lack of a large overlap in the sotas of the average engineering student, engineering professor and practicing engineer as a fourth example of the use of the sota as a collection of heuristics. Presumably the goal of engineering education is to produce an individual who will perform satisfactorily as a practicing engineer. Operationally this goal implies a change in the sota of incoming freshmen to one that overlaps the sota of the engineer in the field. This change is difficult to achieve for two reasons. First, the environment that shapes the sota of the engineer and the one that shapes the sota of the student are different. The cost of failure for a practicing engineer can be quite high; the cost for a student is intentionally limited. In addition, a real design problem may take years to complete, and it may have a large budget, while the student is usually limited to a one-semester design course with no budget at all. Of necessity, the engineer and student work in different environments, and their sotas will evolve differently. Second, the sota of an engineering professor is not the same as that of a practicing engineer. Often a professor has never solved a real engineering problem and has little notion of how this should be done. He is therefore reduced to teaching the theoretical formulas used in design instead of engineering design itself. Not unexpectedly, the result of these two factors is a noticeable difference in the sotas of the graduating senior and the practicing engineer.

Engineering educators have had to develop heuristics to deal with these problems. The traditional approach is to encourage the practicing engineer to participate in engineering education as a guest lecturer and to encourage the professor to take a sabbatical year or consult in industry. Some colleges have also developed design courses that require students to solve authentic problems generated by industry, and others have encouraged students to alternate their formal study with work periods in a cooperative arrangement with industry. All these remedies have merit, but focusing attention on the specific set of heuristics the graduating engineer wraps in his diploma suggests another approach to increase the intersection.

The sota of the graduating student must contain heuristics that allow him to efficiently increase the intersection *after* graduation. While in school the student must learn that an engineering design is defined by its resources and, once in industry, be alert to the heuristics used in resource management. He must also realize that engineering requires that decisions be made amid uncertainty and look for the heuristics the practicing engineer uses to control the risk resulting from this lack of knowledge. Engineering education must not limit itself to trying to achieve an overlap in the sotas of student and engineer at graduation, but must also teach the novice engineer to absorb quickly those heuristics that cannot be taught in school once he is in the industrial environment. The concept of a set of engineering heuristics, or sota, allows the engineering educator to define the goals of modern education and to develop strategies to achieve them.

THE ENGINEER AND SOCIETY:

The relationship between the engineer and society is the last, and most extensive example we will consider of the use of various sotas. It is also one of the most important.

All heuristics are not engineering heuristics, and all sotas are not engineering sotas. As has been observed by other authors, aphorisms, which have all of the signatures of a heuristic, are society's rules of thumb for successful living. Too many cooks do not always guarantee that the broth will be spoiled. And what are we to make of the conflicting advice, "Look before you leap" and "He who hesitates is lost"? As with conflicting heuristics, other rules of thumb in the total context select the appropriate aphorism for use in a specific case. These pithy statements are also dated. Recently the sayings, "There's no free lunch," "Everything is connected to everything," and "If you're not part of the solution, you're part of the problem," have appeared. After a decade and a half, the author of "Never trust anyone over thirty" is publicly having second thoughts about his contribution. Aphorisms are social heuristics that encapsulate human experience to aid in the uncertain business of life.

Society solves problems, society uses heuristics, society has a sota. Some of the heuristics used by the engineer and non-engineer are the same, but each reserves some for exclusive use. Few engineers use a Ouija board, astrology or the *I Ching* in their work, but some members of society evidently do. On the other hand, no layperson uses the Colburn relation to calculate heat transfer coefficients, but some engineers most certainly do. Therefore, the $\text{sota}\,|_{\text{society}}$ and the $\text{sota}\,|_{\text{engineer}}$ are not the same, but will have an intersection as shown in

Figure 7. Here the stippled sota $|_{\text{engineering}}$ from Figure 6 has been combined with a crosshatched one representing society. A heuristic earns admission to the small rectangle of intersection by being a heuristic known to both the engineer and the non-engineer. Figure 7 also includes six solid circles labeled 1 through 6 to indicate subsets of heuristics needed to solve specific kinds of problems. Problem one, lying outside of the sota of society, requires only engineering

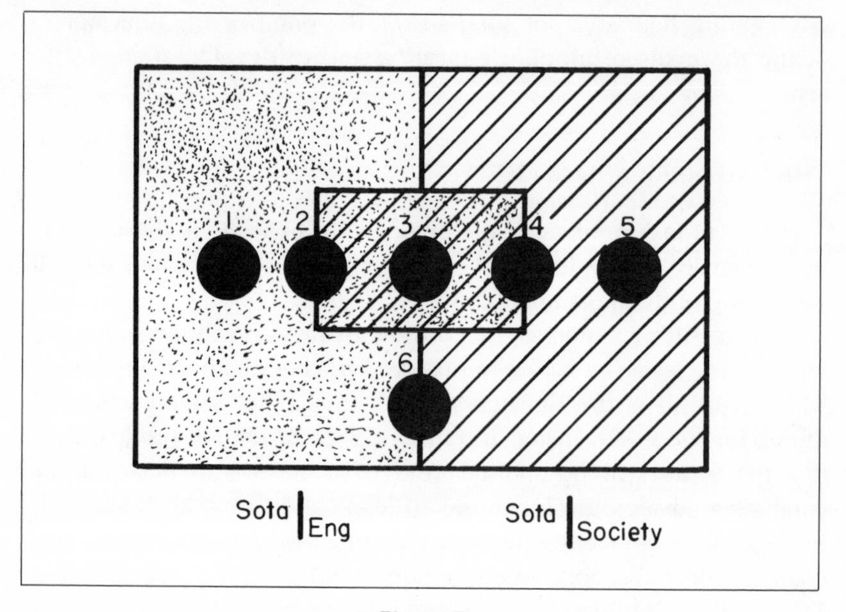

Figure 7

heuristics; problem two requires some heuristics unique to the engineer combined with some from the overlap, and so on. As before, both sota $|_{\text{engineering}}$ and sota $|_{\text{society}}$ are composites of overlapping sotas of individuals, and therefore the problems represented by the subsets 1 through 6 will often require a team effort, possibly including both engineer and non-engineer. Each of these problem areas will now be considered.

Problem one requires only engineering heuristics for its solution. Since the engineer has traditionally responded to the needs of society, few engineering problems lie in this area.

Problems from area two are endemic on the engineer's drafting board. They require information from society to define the problem for solution and heuristics exclusively known to the engineer to solve them. The analysis of the engineer's notion of best given at the end of Part I serves to indicate the importance of this region and the interplay of engineer and society. Non-engineers can never completely understand the trade-offs necessary for the design of an automobile. They must delegate responsibility to the engineer to act in their stead and then trust the engineer's judgment. The alternative approach of restricting the engineer to problems that require no specialized knowledge—that is, those requiring no complicated computer models, no advanced mathematics and no difficult empirical correlations—would soon grind the machinery of engineering to a halt. Problems such as number two require a joint effort in defining goals and solution strategies, but they also require heuristics unique to the trained engineer for their solution.

Some argue that we are witnessing a shrinking of this area as society disciplines the engineering profession because of disagreement over past solutions. No longer is it sufficient for an engineer to assert that a mass transportation system or nuclear reactor is needed and safe. For a problem such as number two, confidence to accept the engineer's judgment outside of the area of overlap is based, in part, on an evaluation of the engineer's performance within the area of overlap, which depends, in turn, on society's understanding of the engineering method. A simple test is in order. Ask the next non-engineer you meet: What does *best* mean to an engineer? How is it related to optimization theory? What is the state-of-the-art? And what is technical feasibility? Those unable to respond satisfactorily to these questions are technologically illiterate and in no position either to delegate important aspects of their life to the engineer or, more important, to discipline the engineer if they do not agree with the engineer's proposed solution. Given the large number of problems in region two and their importance, can society afford humanists who do not have even a superficial knowledge of the major ideas that permeate engineering? What is most urgently needed is research to determine the minimum overlap necessary for a non-engineer to be technologically literate.

Problem three will not detain us long. In this area of complete overlap between society and the engineer, the only dispute is over which heuristics are best to define and solve it: those common to both the engineer and non-engineer, those used by the engineer alone, or those used by one of the various subgroups of non-engineers. The politician, economist, behavioral psychologist, artist, theologian and the engineer often emphasize different aspects of a problem and suggest different approaches to its solution. Nothing could be more different than the heuristics *prayer, positive reinforcement,* and *Freudian psychology* when it comes to rearing a child, or the heuristics used by the politician, economist and engineer when it comes to reducing hunger in the world. Each of us speaks of a problem with the accent of his sota. Many options are available in area three; society's sota must contain effective heuristics to arbitrate between them.

Problem four is distinguished from previous ones in that some of the heuristics needed for an acceptable solution are not found within the sota usually attributed to the engineer in his role as engineer. Remember the San Francisco Embarcadero. Too expensive to tear down, it stands as a monument to a purely engineering solution that failed because the sota used by the engineers did not contain all the heuristics that were important to society. Numerous studies show that compared with the population as a whole, the American engineer is less well-read, a better family member, more conservative politically, more oriented to the use of numbers rather than general philosophical positions in making a decision, and more goal-oriented. These characteristics may change in the future, of course, and are surely different in different cultures, but the conclusion is inescapable. An engineer is not an average person. Accordingly, when he chooses the important aspects of a problem and their relative importance, at times his model will not adequately represent society. The engineer may feel it is obvious that there is an energy shortage and that we need nuclear power; some members of society do not agree. The engineer may feel it is obvious that the scientific view of the world is true; some theologians do not agree.

Only two ways of solving problems in area four are possible. Either the engineer can delegate responsibility for certain aspects of a design to the lay person and accept his input no matter how unreasonable it seems, or he can increase his general sensitivity to the hopes and dreams of the human species—that is, increase the overlap of his sota with that of society. But sensitizing you and me to the human condition is the responsibility of the novelist, psychologist, artist, sociologist and historian—in short, the humanist and the social scientist. Another test is in order. Ask the next engineer you meet: What is the central thesis of behaviorism? What is the difference between a

Greek and Shakespearean tragedy? An engineer unable to respond satisfactorily to these simple questions is illiterate in the liberal arts and in no position either to delegate important aspects of his life to the humanist or, more important, to discipline the humanist if he does not agree with the humanist's proposed solution. Given the large number of problems in region four and their importance, how can society afford engineers who lack even a superficial knowledge of the major ideas that permeate the liberal arts? What is most urgently needed is research to determine the minimum overlap necessary for an engineer to be liberally educated.

Problem five requires heuristics completely outside the expertise of the engineer and is beyond the scope of this discussion.

Finally, problem six is included to give equal time to an aberrant view of engineering. Some people, including some engineers, believe that no overlap should exist between the sotas of society and the engineer. In this view, the duty of society is to pose the problems it wants solved, and the duty of the engineer is to solve them using the best techniques available. This view fails because problems evidently exist that cannot even be defined by society without knowing the range of the technically feasible and because solutions evidently exist that cannot be found without knowing a society's value system. I therefore do not believe that many, if any, examples of problem six exist, or if they exist, that they would be solvable. In spite of its obvious flaws, this limited, technical view of engineering is not as rare as it should be.

Figure 7 underscores the effectiveness of the engineer's concept of a sota in the analysis of the relationship between the engineer and society. It must complete the examples intended to show the value of a sota as a tool for bringing understanding to important aspects of the engineering world. With the heuristic, the engineering method and the sota defined, the discussion returns to our major goal, the rule for implementing the engineering method.

Principal Rule of the Engineering Method

Defining a method does not tell how it is to be used. We now seek a rule to implement the engineering method. Since every specific implementation of the engineering method is completely defined by the heuristic it uses, this quest is reduced to finding a heuristic that will tell the individual engineer what to do and when to do it.

Remembering that everything in engineering is heuristic, no matter how clearly and distinctly it may appear otherwise, I have found I have a sufficient number of rules to implement the engineering method with only one, provided that I make a firm and unalterable resolution not to violate it even in a single instance.

My Rule of Engineering is in every instance to choose the heuristic for use from what my personal sota takes to be the sota representing the best engineering practice at the time I am required to choose.

Careful consideration of this rule shows that the engineer normalizes his actions against his personal perception of what constitutes engineering's best world instead of against an absolute or an eternal or a necessary reality. The engineer does what he feels is most appropriate measured against this norm. In addition to implementing the engineering method, this Rule of Engineering determines the minimum subset of heuristics needed to define the engineer. Recall that in Figure 5 the sotas of three engineers, *A, B* and *N,* overlapped in a small rectangular subset that included the heuristics they shared. If instead of three engineers, all engineers in all cultures and all ages are considered, the overlap would contain those heuristics absolutely essential to define a person as an engineer.

This intersection will contain only one heuristic, and this heuristic is the rule just given for implementing the engineering method. While the overlap of all modern engineers' sotas would probably include mathematics and thermodynamics, the sotas of the earliest engineers and craftsmen did not. While the sotas of some primitive swordsmiths included the heuristic that a sword should be plunged through the belly of a slave to complete its fabrication, the sotas of modern manufacturers of épées do not. The Rule of Engineering is: Do what you think represents best practice at the time you must decide, and only this rule must be present. With that exception, neither the engineering method nor its implementation prejudices what the sota of an individual must contain for him to be called an engineer.

The goal of Part I of this discussion was to describe the situation that calls for an engineer. The goal in Part II has been to describe how the engineer responds when he encounters such a situation. If you desire change; if this change is to be the best available; if the situation is complex and poorly understood; and if the solution is constrained by limited resources, then you too are in the presence of an engineering problem. What human has not been in this situation?

If you cause this change by using the heuristics that you think represent the best available, then you too are an engineer. What alternative is there to that? To be human is to be an engineer.

The definition of the engineering method depends on the heuristic; the rule of the method and the Rule of Judgment are heuristics; and the engineer is defined by a heuristic—all engineering is heuristic.

PART III
SOME HEURISTICS USED BY THE ENGINEERING METHOD

"Nothing of any value can be said on method except through example," counsels the eminent philosopher, Bertrand Russell. Cowed by such an authoritative rule of thumb, the first objective of Part III will be to sample the smorgasbord of engineering heuristics. Since examination of a long list quickly satiates, these heuristics have been collected into five categories. The division is arbitrary and only for ease of reference. A definitive taxonomy of engineering heuristics must await another forum. Grouped together are:

1) Some simple rules of thumb and orders of magnitude,
2) Some factors of safety,
3) Some heuristics that determine the engineer's attitude toward his work,
4) Some heuristics that engineers use to keep risk within acceptable bounds, and
5) Some rules of thumb that are important in resource allocation.

The reason for this extensive review is to insist on the broad meaning I intended to give the word *heuristic* in the assertion made at the end of Part II, that "all engineering is heuristic." To this end, several examples will be cited for each of the above categories. The specific examples chosen are not important in themselves and many others would serve as well. Their large number and wide variety are important, however, in establishing the scope of the engineering heuristic. By its extent, this list will distinguish my view from that of recent authors who limit the engineering heuristic to a routine adaptation of its traditional role in problem solution.

The second objective of Part III is to examine competing definitions of the engineering method. What we will find is that none of these alternative definitions is absolute, and each is therefore appropriately included in this section as an additional engineering heuristic.

The third and final objective of Part III is to reexamine the definition of the engineering method given at the end of Part II in light of the progress we have made and to put it in final, compact form.

Sample Engineering Heuristics

I do not know whether I ought to touch on the simplest heuristics used by the engineer, for they are so specific to the problem they are intended to solve that they are often unintelligible even to engineers in closely related specialties and hence may not be of interest to most people. Nevertheless, to test whether my fundamental notions are accurate and to whet the appetite for what is to come, I feel more or less constrained to speak of them. In listing these simple heuristics I do not intend to instruct in their use or even to reach an understanding of what they mean, but rather to establish their existence, their variety, their number and their specificity.

Rules of Thumb and Orders of Magnitude

In engineering practice, the terms *rule of thumb* and *order of magnitude* are closely related, often used interchangeably and usually reserved for the simplest heuristics. My colleague who estimated the size of a room knowing the order of magnitude for standard column spacing was using the kind of heuristic I have in mind, as is the civil engineer who quickly estimates the cost of a proposed highway by remembering the rule of thumb that a typical highway in America costs one million dollars per mile. Both an order of magnitude and a simple rule of thumb must be considered as heuristics, of course, because neither guarantees the correct answer to a problem. Highways costing more or less than one million dollars per mile certainly exist, and given the sotas of some avant-garde architectural engineers, I would not be surprised to find buildings somewhere with irregular column spacing. Still, both are useful to the practicing engineer whose work would be severely handicapped were these simple heuristics to become unavailable tomorrow.

These two examples demonstrate the existence and, by implication, the importance of simple rules of thumb, but they give little indication of their variety. The following group of heuristics, chosen at random from the various branches of engineering, will correct any chance misimpression. The

> ***Heuristic*: The yield strength of a material is equal to a 0.02 percent offset on the stress-strain curve**

is used almost universally by mechanical engineers to estimate the point of failure of a wide variety of materials, and the

> ***Heuristic*: One gram of uranium gives one megawatt day of energy**

is needed by the nuclear engineer for a quick-and-dirty estimate of the amount of energy a power plant will generate. The chemical engineer making heat transfer calculations often assumes the

> ***Heuristic*: Air has an ambient temperature of 20° centigrade and a composition of 80 percent nitrogen and 20 percent oxygen.**

when, in fact, the chemical plant he is designing may be located on a mountain where this rule of thumb is not exact but only an approximation. Similarly, today the

> ***Heuristic*: A properly designed bolt should have at least one and one-half turns in the threads**

may appear banal, but its continued use proves its continued value. This list could go on at length. As it stands, however, it is sufficient to emphasize the wide variety of engineering orders of magnitude and to demonstrate the hopelessness of trying to compile a complete list of heuristics used by any one engineer, much less the engineering profession.

The engineer uses hundreds of these simple heuristics in his work, and the set he uses is a fingerprint that uniquely identifies him. The mechanical engineer knows the importance of the 0.02 percent offset on the stress-strain curve and the number of turns on a properly designed bolt, but probably has no idea how to estimate the energy release in a nuclear reactor. The chemical engineer knows the number of plates in the average distillation tower, but does not know the strength of concrete or the average span of a suspension bridge. An engineer's simple rules of thumb and orders of magnitude are sufficient to identify his discipline, culture and education. They are the ammunition each engineer uses in his own private preserve.

These simple rules of thumb and orders of magnitude represent the first category of engineering heuristics.

Factors of Safety

One type of simple heuristic is so valuable that it is isolated here for special consideration. I am referring to the engineering numbers called *factors of safety*. When an engineer calculates, say, the strength of a beam, the reliability of a motor or the capacity of a life-support system, approximations, uncertainties and inaccuracies inevitably creep in. The calculated value is multiplied by the factor of safety to obtain the value used in actual construction. If anyone still doubts that engineers deal in heuristics, the almost universal use of factors of safety at all steps in the design process should dissuade him from that notion. In the factor of safety we see the heuristic in its purest form—it does not guarantee an answer, it competes with other possible values, it reduces the effort needed to obtain a satisfactory answer to a problem and it depends on time and context for its choice. An example will make this concept clear.

The evaluation of the wall thickness of a pressure vessel requires many heuristics. At times these will include mathematical equations, handbook values, complex computer programs and laboratory research. None of these gives an exact answer. To quote one of my former chemical engineering professors, "Always remember that experimentally determined physical properties such as viscosity and thermal conductivity are evaluated in the laboratory under pristine conditions. In the actual vat where the stuff is manufactured, you will be lucky if someone has not left behind an old tire or automobile jack." Uncertainty in the calculated value is always present and no experienced engineer would ever believe that the above-mentioned heuristics could produce an absolutely correct value for the wall thickness of a pressure vessel. To compensate for this uncertainty, he will multiply the answer he calculates by a factor of safety. Instead of using a calculated thickness of eight inches, he may, for example, prescribe one of ten, twelve or even sixteen inches. In this way, many serious problems, due to the inherent uncertainty of the engineering method, are never allowed to develop.

Attitude-Determining Heuristics

Knowledge of the rules of thumb we have just noted and others like them distinguish the engineer from the non-engineer, but this is not the only difference between the two. Our present interest should not be limited to technical examples to find such a distinction, but should also focus on those heuristics that define the attitude or behavior of an engineer when he is confronted with a problem. What does he do? How does he act? What heuristics determine the engineer's attitude toward his work and establish his unique view of the world? Our sample will include two heuristics as representative of the category: the engineer's mandate to give an answer when asked and his determination to work at the margin of solvable problems. Although some of these examples have been hinted at before, they are repeated here to demonstrate a group of heuristics that are not directed specifically at seeking the solution to a problem but are still essential and very much a part of the engineer's approach to problem solution.

The willingness to decide or the willingness to give an answer to a question, any question, is an example of the proper engineering attitude. The more original and peculiar the question, the more evident the distinction between the engineer and the rest of the population. The student willing to estimate the number of ping-pong balls that could be put into the classroom was obeying the engineering

> ***Heuristic:* Always give an answer.**

This heuristic is often taught explicitly to engineering students. For example, the design of distillation towers, those familiar tall towers that dot the landscape of a chemical plant to refine petroleum products, involves the calculation of the number of plates or stages they should contain. The theoretical analysis, whose exact nature is of no concern to us now, requires a graph called a McCabe-Thiele diagram. One of my former professors once told our class in a stern voice, "If you are ever in the board room of a large chemical company and are asked for the number of distillation plates needed to distill a material with which you are unfamiliar, guess thirteen. I'm here to tell you that as a good rule of thumb, the average number of plates in distillation towers in the United States is thirteen.* If you know something about the McCabe-Thiele diagram for the substance in question, perhaps that it has a bump here or a bulge there, up your estimate by ten percent or lower it by the same amount. But if you admit that you have been in my class in distillation, for heaven's sake don't say 'I don't know.'"

Although this is a true story, it is perhaps an exaggeration. Its point, however, is clear. Engineers give the best answer they can to any question they are asked. Of course in answering, the engineer assumes that the person asking the question is literate in the rules of technology and understands that the answer provided is in no sense absolute but rather the best available based on some commonly acknowledged sota.

Characterizing engineering design as the use of engineering heuristics implies that the attitude of the engineer is controlled by the additional

***Heuristic*: Work at the margin of solvable problems.**

Neither problems amenable to routine analysis nor those beyond the reach of the most powerful existing engineering heuristics are included in what may properly be called engineering. An algebra problem requiring only known, presumably uncontroversial, rules of mathematics certainly would not be called an engineering design problem. On the other hand, a problem completely beyond the reach of even the most powerful engineering heuristics, one well outside of the sota of the engineer, would also be disqualified. Engineering design, as traditionally conceived, has no heuristics to answer the questions: What is knowledge? What is being? What is life? To qualify as design, a problem must carry the nuance of creativity, of stepping precariously from the known into the unknown, but without completely losing touch with the established view of reality. This step requires the heuristic, the rule of thumb, the best guess. If it were possible to plot all problems on a line from the most trivial to the most speculative, the engineer uses heuristics to extrapolate along this line from the clearly solvable problems into the region where the

* Now some twenty years later in 1982 (the time stamp, once again) a better number is probably 20.

almost or partially solvable problems are found. He works at the margin of solvable problems.

We have noted that the engineer is different from other people. His attitude when confronted with a problem is not the same as the average person's. The engineer is more inclined to give an answer when asked and to attempt to solve problems that are marginally solvable. These examples complete the selection of typical heuristics that show the engineer's attitude toward problem solution. It does not, however, include all of those that could be considered or even the most important. The engineer is also generally optimistic, convinced that a problem can be solved if no one has proved otherwise, and willing to contribute to a small part of a large project as a team member and receive only anonymous glory. The heuristics mentioned here are sufficient, I think, to indicate the presence of heuristics in the engineer's sota beyond those traditionally associated with problem solution. Any serious effort to explain the engineering method must account for these heuristics that define the engineer's attitude when confronted by a problem.

Risk-Controlling Heuristics

Because the engineer will try to give the best answer, even in situations that are marginally decidable, some risk of failure is unavoidable. This does not mean, of course, that all levels of risk are acceptable. As should be expected by now, what is reasonable is determined by additional heuristics that control the size risk an engineer is willing to take. A representative group of these risk-controlling heuristics will be discussed now, including: make small changes in the sota, always give yourself a chance to retreat, and use feedback to stabilize the design process.

The first

> ***Heuristic*: Make small changes in the state-of-the-art,**

is important because it stabilizes the engineering method and explains the engineer's confidence in using contradictory, error-prone heuristics in solving problems, even those involving human life. Since no way exists, in advance, to ensure that a given set of heuristics will produce a satisfactory solution to a given problem, prudent practice dictates using this set only in situations that bear a family resemblance to problems for which a successful solution has been found. In other words, within the hypothetical set of all possible problems, a new problem to be solved heuristically should find itself in or near the cloud of already-solved problems. To illustrate this point, the sets *U* and *E* from a previous example are reproduced in Figure 8.

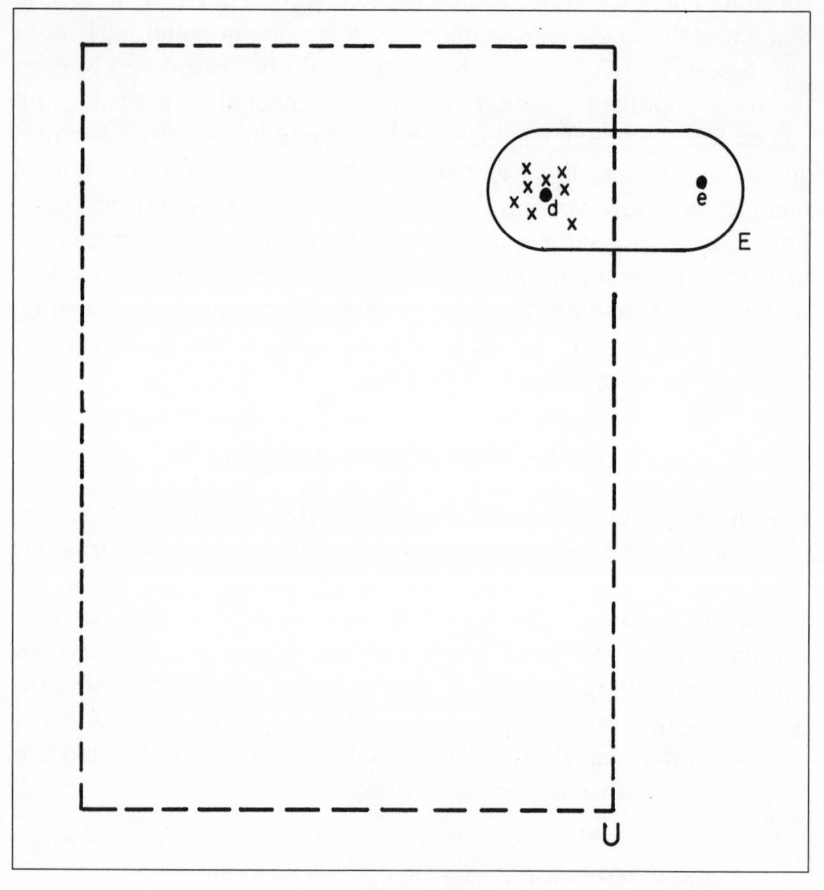

Figure 8

Let us assume that in the past the heuristic *E* has been successfully applied to problem *d* and to other problems, marked with *x*'s, that bear a marked family resemblance to *d.* In effect, the engineer has built up engineering experience with *E.* The engineering heuristic under consideration counsels the engineer to use *E* only when he can apply it to a problem located within the cloud of *x*'s in Figure 8.

Under this condition the engineer is reasonably secure in using fallible, nonanalytic solution techniques. But errors do creep in. The exact position of the dotted boundary in Figure 8 is not known, and occasionally the engineer will stray across it and a design will fail. One of the most spectacular engineering failures was the Tacoma Narrows Bridge in Washington state. By oscillating with increasing amplitude over a period of days before crashing into the river below, it earned the name "Galloping Gertie." When accidents happen, the engineer is quick to retreat, or as he would say, *back off* in the next use of *E*. By failure he has explored the range of validity of his heuristics.

A small step does not imply no step. Progress is made as the engineer navigates from the safety of one bank to the unknown bank on the other side of the stream, using heuristics as his guide. The design of the first chemical plant to produce nylon proceeded from stepping stone to stepping stone as the original theoretical idea became the bench-top experiment, the pilot plant, the demonstration plant and finally the full-scale plant itself. This sequence, under the firm control of heuristics, allowed a safe extrapolation as knowledge gained at one step was passed to the next until the material for the blouse or the shirt you are now wearing could be produced. As with the blind man tapping his way down an unknown path, the engineer makes his way carefully in the darkness. He resolves to go so slowly and circumspectly that even if he does not get ahead very rapidly, he is at least safe from falling too often. In spite of its uncertainty, the heuristic method is an acceptable solution technique in part because of the stabilizing effect of this heuristic in the typical state-of-the-art.

I remember when I first heard the next engineering

> ***Heuristic*: Always give yourself a chance to retreat,**

explicitly stated. It was in a laboratory course offered by a professor who later became a commissioner for the Atomic Energy Commission. I don't remember the specific example he used, but I do remember his point. Much as the computer technician stores the daily operation of his computer on a back-up tape or as a sensible person mentally checks where his house keys are before locking the door behind him, this heuristic recommends that the engineer allocate some of his resources to preparing for an alternative design in case the chosen one proves unworkable. Or, to use one of society's aphorisms, "Don't put all your eggs in one basket."

In a previous section, nuclear engineers were quoted as saying that a fundamental difference in the testing philosophies of American and Russian engineers was that the former tested many design variations before settling on one, while the latter preferred to decide quickly on a reactor type, build it and then try to make it work. Since a reactor has many components (the fuel, coolant, moderator, reflector and shield) and since many choices exist for each of these components, a large number of different reactor types are possible. To name only a few, engineers have designed pressurized water, boiling water, heavy water, homogeneous aqueous, molten plutonium, molten salt and high-temperature gas-cooled reactors. Early in nuclear history, American engineers built a bench-top experiment, pilot plant and demonstration plant for as many of these different types as possible. Russian engineers, on the other hand, selected only a few reactor types early in their nuclear program and allocated their resources to them. The difference in the two programs is shown in Figure 9, with the American system at *A* and the Russian at *R*. At *A*, money is allocated for a preliminary evaluation of all possible reactor types as indicated by the lower level of the pyramid. As the evaluation proceeds, the remaining resources are funneled to the most promising concepts in an ever-narrowing manner as the engineer seeks the one best design. At *R*, the Russian plant calls for a much earlier choice of reactor type to which all resources are allocated. A careful comparison reveals that both heuristics have advantages and neither can be rejected out of hand. If the design engineer can be reasonably certain that his initial choice is near optimum, or that all choices are equally desirable, the Russian system, by requiring fewer resources to reach the design objective, is clearly preferable. It does not, however, offer a chance to retreat. If the chosen reactor type proves physically unrealizable or economically unsound during the design process, as indicated by the X at *R′*, the Russian engineer must begin at the beginning as indicated by the dotted square. The American approach, at *A′*, is more extravagant with resources, but the extra time and money invested in design alternatives can contribute in two ways: first, by assuring that the final design is nearer to the optimum choice and second, by allowing retreat to a lower level if the first choice is blocked. A related formulation of the rule of thumb that an engineer should allow himself a chance to retreat recommends that design decisions that carry a high penalty should be identified early, taken tentatively, and made so as to be reversible to the extent possible. In a complex, unknown system, the possibility of retreat to a solidified information base will often pay dividends.

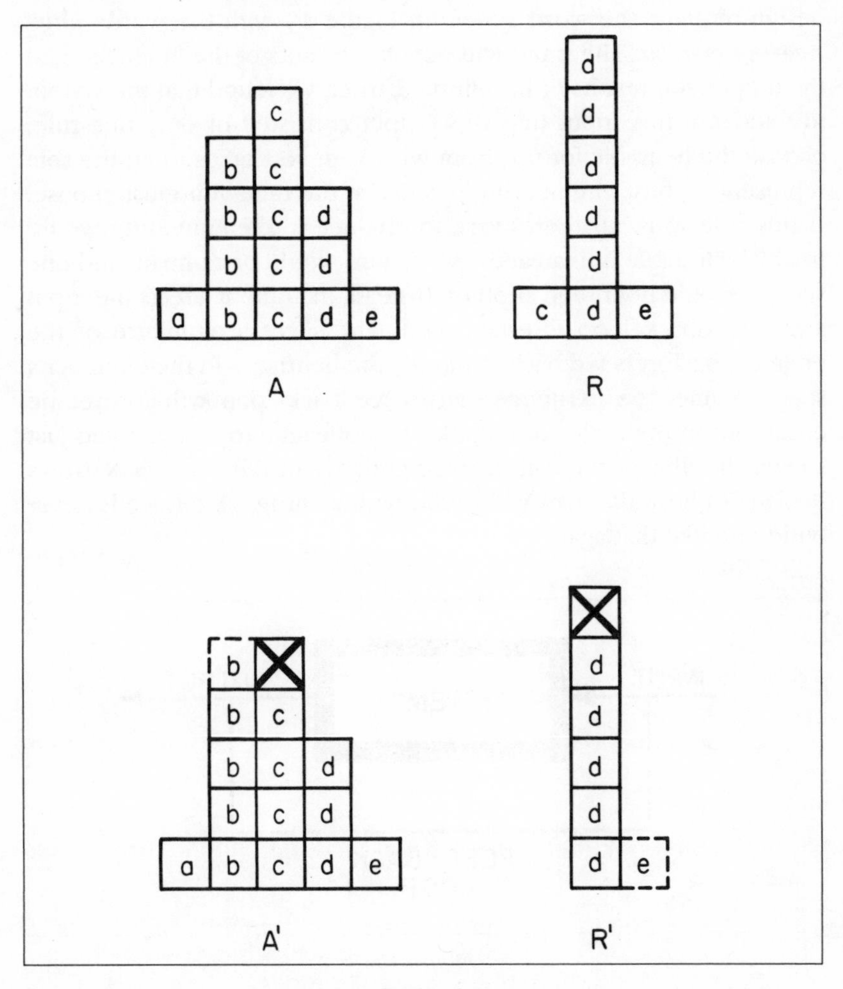

Figure 9

Although the engineering term *feedback*, which is at the heart of the

***Heuristic*: Use feedback to stabilize engineering design**

dates from the water clock of Ktesbios in the third century B.C., the theoretical analysis of feedback is barely fifty years old. Recall that feedback is the arrangement of any system, whether electrical, mechanical or biological, in which the output affects the input.

To see the parallel between a feedback system and engineering design, replace the word *system* in Figure 10 with the words *engineering method.* The input will now be the sota of the engineer, and the output, the results of his efforts. Earlier, we found that the system to transform this input into this output consisted of only one rule: choose the heuristic for use from what your sota takes to be the sota representing best engineering practice at the time you must choose. If this rule were all there were to engineering, engineering would hardly be a stable human activity—failure would be rampant and one success would not breed another. Instead, the output affects the input and a feedback loop is established. The success or failure of the engineer's effort is fed back to modify the heuristics in the engineer's sota. For me, the existence of this feedback loop will forever be enshrined in the sardonic remark of a colleague to whom I had just shown the film of the catastrophic collapse of the Tacoma Narrows Bridge. As he walked away, he said with a shrug, "Well, we'll never build one like that again."

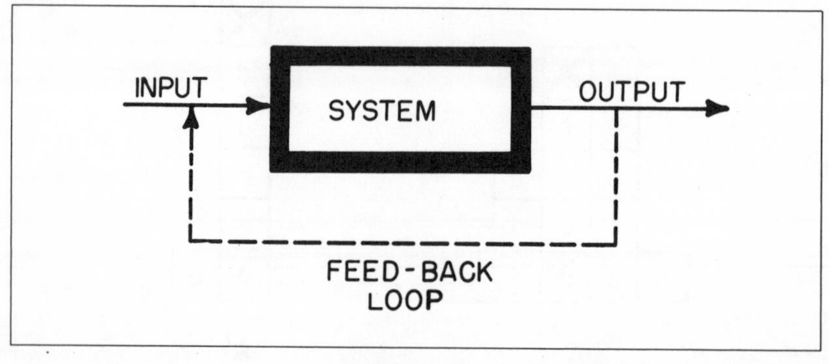

Figure 10

If a bridge falls, films of the failure are studied, models of the bridge are tested in wind tunnels and competing methods of calculation are examined to see which most accurately predicted the problem. As a result, the sota of bridge design changes. Stable engineering design requires this feedback.

Engineering is a risk-taking activity. To control these risks, the engineer has many heuristics in his sota. For example, he makes only small changes in what has worked in the past, tries to arrange matters so that if he is wrong he can retreat, and feeds back past results in order to improve future performance. Any description of engineering that does not acknowledge the importance of these three heuristics and others like them in stabilizing engineering design and, in effect, making engineering possible, is hopelessly inadequate as a definition of the engineering method.

Resource Allocation Heuristics

Since an engineering problem is defined by its resources, I would be remiss if at some point I failed to include a sample of the many heuristics the engineer uses to allocate and manage available resources. This is our fifth category of engineering heuristics. It will be represented by the heuristics: allocate sufficient resources to the weak link; allocate resources as long as the cost of not knowing exceeds the cost of finding out; and at the appropriate point in the project, freeze the design.

The first item on the list, the

> ***Heuristic:*** **Allocate sufficient resources to the weak link,**

ostensibly refers to the English aphorism, "A chain is as strong as its weakest link." By extension, it implies that if a stronger chain is desired, the correct strategy is to strengthen this link. The same concept exists in other disciplines, chemistry for one, where the weak link is now called the limiting reagent. Since one atom of sodium combines with one atom of chlorine to produce one atom of salt, the final amount of salt in this chemical reaction depends on which of the original elements, the limiting reagent, is in short supply. In this, as in all problems of chemical stoichiometry, if you want more product, increase the limiting reagent; in effect, attack the weak link.

Engineering design is no different. With the possible exception of the famous one-hoss shay built in 1755* that was reputed to have had no weak link, every engineering project has a limiting element in its design. Good engineering practice requires that sufficient resources be allocated to this element. The overall design will be no better—and due to overdesign of the less important parts of the final project, may actually be worse—if additional resources are allocated to the less critical components.

*"Deacon's Masterpiece or the One-Hoss Shay" reported by Oliver Wendell Holmes.

This heuristic is so important that in certain limited cases the engineer has produced a theoretical formulation of it, such as in the scheduling and coordination of the many individual tasks needed to complete large engineering projects. Construction of a building, airplane or bridge requires that blueprints be drawn, the site prepared, materials procured and verified, personnel hired and so on. Some of these tasks can be performed in parallel; others must await the completion of an earlier task. The critical path method and the related performance evaluation and review techniques (often called by their engineering acronyms CPM and PERT analysis) are mathematical strategies for scheduling individual tasks and finding the critical sequence of them (or as we would say, the weak link) for completing a project. The overall time to construct a bridge is not shortened if additional resources are allocated to finish tasks not found on the critical path. Theory predicts that if you want to decrease the time to complete the project, attack this weak link.

The second item on the list, the

> ***Heuristic:* Allocate resources as long as the cost of not knowing exceeds the cost of finding out,**

appears frequently in the literature in a variety of forms. One author suggests "that a project be continued when confidence is high enough to permit further allocation of resources for the next phase or should be discontinued when confidence is relatively low." Another asks "if what has been learned about the project to date or the current prospects of yielding a satisfactory answer justify continuing to invest additional resources." A third author prefers the question, "Does what we know now warrant continuing?" Each of these formulations is essentially the same and simply acknowledges the trade-off between knowing and not knowing. In each case, two conflicting options carry an associated cost, and the engineer must decide, heuristically, of course, which cost is lower.

An interesting ramification of this rule of thumb condones the engineer's refusal to explore preposterous design alternatives. A subtle distinction exists between justifying consideration of an alternative view on the basis of the current sota, as is typical of the engineer, and on the basis of its approximation to truth, as is typical of the scientist. Since truth is generally held to be an absolute good (and what scientist would not prefer a theory closer to the truth than one that was not), in principle science must grant all points of view, no matter how bizarre, an audience. In spite of the initial strangeness of the Theory of Relativity, it would have been bad scientific taste to reject it before it had done battle against other theories to prove its mettle against truth.

Not so for an alternative engineering design concept. Some years ago, a crank device called the Dean Machine was proposed to solve the problems of air travel. The device was alleged to be able to hover and fly for an indefinite period without any outside sources of energy or any interaction with a known field. On the face of it, the Dean Machine violates essentially every known law of science beginning with Newton and ending with the Second Law of Thermodynamics. No working model was provided, and the few available schematic diagrams showed a complex arrangement of eccentric cams that would have required many hours to analyze. Brandishing such taunts as "They laughed at Galileo" and "They laughed at Einstein," proponents of the Dean Machine accused the engineering fraternity of being closed-minded, cliquish and afraid of unconventional ideas for not sponsoring research to prove whether the machine would work. Although science has no rule of procedure to dismiss any theory before the battle for truth is joined, engineering does. Since analysis of the Dean Machine would require a large amount of the available resources, since it represents a very large change in the sota and since it could hardly be called best engineering practice, even after only a cursory glance the engineer can justify rejecting further study because "what he now knows does not warrant continuing."

Of course, this heuristic, as with all others, does not guarantee that the absolute best decision has been made. Considering the large penalty associated with resources squandered on a wild goose chase, if the available resources restrict the engineer to the well-traveled path, he need not regret the opportunity missed on the road not taken.

I cannot certify from personal experience the observation of the English engineer who writes:

> Rightly or wrongly, the U.S.A. has the reputation of being able to develop a new invention much more readily than we do in this country. If this is true, it may well be that one of the reasons for it is that the Americans usually veto any improvement in design after construction has begun. Leave it alone and alter the design in the next machine or the next batch; don't tinker with this one is their policy. And it is a highly realistic one.*

*Glegg, *The Science of Design*, Cambridge University Press, 1983.

If this statement is accurate, it expresses a significant difference in the sotas of two countries with respect to the

> ***Heuristic:*** **At some point in the project, freeze the design**

for this heuristic is quite common in the sota of the American engineer. Occasionally, it is even explicitly expressed, as in one recent book on fast nuclear reactor theory that reminds the reader that the set of numbers it gives was fixed by the size of the computer code when "the design was *frozen.*" This rule of thumb recognizes that a point is often reached in design where the character of a project, and hence the appropriate allocation of resources, changes rather brutally from seeking alternative solutions to one of perfecting a chosen solution. As might be expected, this point is located heuristically by a trade-off between the relative risk and benefit of seeking yet another alternative. After it is reached, a major design change runs too large a risk of introducing a fatal flaw, because insufficient resources remain to evaluate all its ramifications. Once a design has been frozen (as a good rule of thumb, about 75 percent of the way into the project), the members of the design team take the general attitude, "Let's go with it."

The three heuristics—attack the weak link; allocate resources as long as the cost of not knowing exceeds the cost of finding out; and at the appropriate point, freeze the design—are not the only resource allocation heuristics in the armory of the engineer. They are, however, excellent examples of this important class and worthy of special study.

We have now completed the first major objective of Part III by sampling a few simple rules of thumb, orders of magnitude and factors of safety as appetizers, and by avoiding the temptation to spend too much time on the heuristics that determine an engineer's attitude toward his work, those that control the risks he takes and those that help him allocate the available resources. What we have found is that the range of engineering heuristics is much broader than usually recognized. With more time, we could easily extend this range even further. Any serious attempt to define the engineering method must, at the very least, account for all the heuristics we have just seen.

Alternative Definitions of Engineering

The second objective of Part III is to give equal time to the opposition. We need to examine alternative definitions of the engineering method and show why each falls short of an absolute definition and should, itself, be taken as a heuristic. This study will not take long, for unlike the extensive efforts to define the scientific method, the need to define the engineering method in a philosophically justifiable way has not been critically felt until recently. Four competitors will be examined: the definition of the engineering method as: 1) adherence to a specified morphology, 2) applied science, 3) trial and error, and 4) a problem-solving, goal-directed or needs-fulfilling activity.

Engineering and Morphology

The most common and ambitious effort to define the engineering method is the attempt to associate it with a specific, universal structure, the so-called morphology of engineering design. Many authors have tried to define the engineering method by listing a fixed sequence of steps through which the design process is assumed to pass. For example, one recent effort gives the structure of design as: analysis, synthesis and evaluation. That is, the engineer—and if we seek a definition, presumably *only* the engineer—

1) Analyzes a problem,
2) Synthesizes a solution, and
3) Evaluates the results.

A more classic morphology directed at problem solution in general, but adapted frequently to engineering, is: understand, plan, carry out and examine. By which is meant, the engineer must

1) Understand the problem,
2) Devise a plan to solve the problem,
3) Carry out the plan and, finally,
4) Look back to check the solution obtained.

Perhaps the most extreme example of a morphology is that of the author who insists that to be called an engineer you must

1) Determine the specifications,
2) Make a feasibility study,
3) Perform a patent search,
4) Develop alternative design concepts,
5) Determine the selection criteria,
6) Select the most promising design concept,
7) Develop a mathematical or physical model,
8) Determine the relationship among the basic dimensions and materials of the product,

9) Optimize the design,
10) Evaluate the optimized design by extensive analysis on the mathematical model and tests on physical models and, finally,
11) Communicate the design decisions to engineering administrative and manufacturing personnel.

The basic assertion in each of these proposed definitions of the engineering method is that we know what an engineer is, what he does and how he does it, if we can produce a list of steps in a fixed order that must be followed to produce a product that is identified as the result of engineering.

Although these proposed structures are often helpful as heuristics, and the components of each structure often reveal many important heuristics used by engineers, structure is inadequate as a definition of design for four reasons. First, while many of the proposed structures vaguely resemble each other, most are the eccentric vision of their author. Pick up any recent book on engineering design to see the currently popular list. Between the two extremes just given, I found 25 variations on the theme before I stopped counting. From a practical point of view, a rule of thumb is needed to choose from this variety. Introducing a heuristic at this point reduces a question of dogma to one of style.

Second, the more candid authors admit that engineers cannot simply work their way down a list of steps but must circulate freely within the proposed plan—iterating, backtracking and skipping stages almost at random. Soon structure degenerates into a set of heuristics badly in need of other heuristics to tell what to do when.

Third, none of the structures proposed so far recognizes the full spectrum of heuristics essential to a proper definition of the engineering method. Where are we counseled to make small changes in the sota? To allocate resources to the weak link? To use simple rules of thumb? The essence of engineering is not captured in the commands: analyze, synthesize and evaluate.

Finally, to paraphrase what a scientist once said of efforts to define the scientific method as a sequence of steps, the fourth reason why structure is inadequate as a definition of the engineering method is that in actual practice it is highly unlikely that engineers follow any structure proposed to explain their work. Do we really believe that Neanderthals, primitive artisans, early engineers or even a team of modern engineers proceeded by first completely understanding their problem, then completely developing a plan, next completely carrying out this plan and finally examining completely the solution obtained? If we do not believe that they did so, are we sure that we have the right to disfranchise them from engineering and say they are not engineers? Therefore, from the perspective of this discussion, a morphology is a set of heuristics, a specific sota. It is useful as a heuristic for the novice engineer, but insufficient as an absolute definition of the engineering method. As a result, we are left with the

> ***Heuristic:*** **Use a morphology to solve engineering problems.**

Engineering and Applied Science

Some authors, primarily those with limited technical training, incorrectly assert that engineering is applied science. This is the second most common attempt to define the engineering method. Misunderstanding the art of engineering, these authors become mesmerized by the admittedly extensive and productive use made of science by engineers and elevate it from the status of one valuable engineering heuristic among many others to identity with engineering itself. On careful analysis, however, the engineer recognizes both science and its use as heuristics, although very important ones, to be applied only when appropriate.

The thesis that engineering is applied science fails because scientific knowledge has not always been available and is not always available now, and because, even if available, it is not always appropriate for use.

Science, using the word anywhere near its present connotations, is a relatively new human invention. Most historians credit the Ionian natural philsophers of the sixth century B.C. as its founders. When Thales taught that everything was made of water and his disciple, Anaximander, disagreed, the human species saw the birth of what has variously been called the Greek way of thinking, the comprehensibility assumption and the scientific myth. Not surprisingly, I call it the Greek heuristic. It is instructive to remember that those cultures derived from ancient China and informally based on what might be called the "Chinese way of thinking" do not endorse this heuristic. Other cultures, such as the French, whose once strong rational tradition is now greatly weakened by modern philosophy under the influence of Martin Heidegger and, perhaps to a certain extent, Henri

Bergson, are retreating from it. To be sure, some historians have found traces of doubt and criticism before the sixth century B.C., and others have found Asiatic anticipations and variants of this same approach to explanation in Indian and Chinese literature, but the definitive hypothesis that the world is comprehensible and accessible to critical analysis was first made by the sages of the Milesian school in Ionia. This assumption is therefore a Greek invention—one of the pivotal inventions in human history. So pervasive has this hypothesis become that one historian has been led to assert that "it is an adequate description of science to say that it is thinking of the world in the Greek way."

Of course, before the beginning of systematic inquiry, humanity had acquired a jumble of ideas about the world that was sufficient for gathering food, constructing shelter and managing daily affairs. But these beliefs were characterized by superstition, imprecision, contradiction, lack of knowledge of their range of application, confused interrelationships, appeals to mystical forces and dependence on custom, rather than truth, for their certification. In a word, they were heuristics. As a body they defined a sota that was crassly utilitarian and tuned to answer the questions of the moment. This sota was sufficient, after a fashion, for building bridges, irrigation canals, dams, homes—and sepulchers. In it was to be found nascent engineering. Engineering, the use of engineering heuristics, clearly predates science, the assumption that the world is amenable to critical analysis. With science yet to be discovered, early engineering could hardly be defined as applied science.

What of the present day? Scientific knowledge is still unavailable for some, perhaps most, of the decisions made by modern engineers. The design of a system to put a man on the moon could not have depended exclusively on applied science, because no one had ever been to the moon before and could not know precisely what science to apply. The exact temperature, pressure, gravitational field and composition of the moon were unknown. Without science, how do you apply it? Yet a man placed his foot on the moon on July 20, 1969.

A second reason that engineering cannot appropriately be called applied science is that sometimes the engineer does not use available scientific knowledge that bears on his problem. Since an engineering problem is defined by its resources, an engineer must make his decisions within the amount allocated. Developing, retrieving and applying scientific knowledge always incurs cost. In some cases the engineer is so poor that he can afford only past experience, intuition, folklore and educated guesses to solve his problems, while in others he is rich enough to afford science. The choice in each instance is dictated by his sota. It is simply not the case that the engineer uses science when available regardless of the cost. Whether because science is unavailable or because it is too expensive, the thesis that engineering is applied science must be rejected. We must admit that modern science, as a heuristic, has fueled the machinery of modern engineering, but we should not assume it is the machinery itself.

In spite of the philosopher's problem in determining the correct epistemological status of science, the engineer perceives no serious difficulty, for he requires science only as the

***Heuristic:* Apply science when appropriate.**

Engineering and Trial and Error

The problem solution strategy called trial and error has had a curious history in engineering. Undoubtedly it was first encountered as a technique for solving complex problems. Even well-posed problems (in the sense that a sufficient number of independent relationships exists between the variables to ensure a unique answer) often frustrate a mechanical solution procedure because some of the necessary information is available only in graphical, transcendental or tabular form. In each of these cases, guessing the final answer (a trial) and then verifying that it is correct (not an error) is often the simplest way to proceed. This strategy is fairly common in some branches of engineering, notably chemical engineering, where the trial-and-error solution of embedded equations is almost a way of life.

Engineering does not, however, reduce to a simple trial-and-error procedure. In engineering, a wide variety of designs is not tried randomly, then measured against an absolute set point, after which the failures are eliminated and the most successful retained. The problem with this analysis is that, if anything, the engineering prophet is too good. The ratio of engineering successes to failures is unexpectedly high. No matter how difficult an engineering task appears, somehow it always succumbs. The engineering goals of designing a supersonic airplane capable of flying faster than Mach 2, of landing a man on the moon and returning him safely to earth and of building a power plant to exploit the nuclear fission reaction have all been established. Now supersonic airplanes, moon landings and nuclear fission reactors all exist. Any explanation of the engineering method must explain this high success rate. Of course, a few engineering projects do fail, but these failures are always greeted with surprise. By and large, engineers are too successful at everything they try for simple, random trial and error to be the answer.

Instead, as we saw earlier, information derived from the completed project is returned or fed back to modify the structure of the engineering sota in a fundamental way. If past designs did not affect present designs directly and essentially, not only Gertie but the majority of her progeny would gallop.

Any phrase that feigns to explain the engineering of the present day must, at the least, be powerful enough to whisk us safely from the take-off of a small aircraft at Kitty Hawk on December 17, 1903, to a landing in the Sea of Tranquillity on July 20, 1969, with relatively few crashes. Only the identification of the engineering method with the use of engineering heuristics—where one of the included engineering heuristics is "use feedback to stabilize engineering design," not identification of the engineering method as "trial and error"—can do so. The

***Heuristic:* Engineering is trial and error**

is simply inadequate as a definition of the engineering method.

Engineering and Problem Solution

Many other definitions of the engineering method exist. In conventional practice they are rather vaguely associated with either solving a problem, attaining a goal or fulfilling a need. Thus we read: "design is a goal-directed, problem-solving activity"; "design is a creative decision-making process directed toward the fulfillment of human needs"; and so on. These definitions are convenient and, when speaking informally, I have used them myself. But either because they raise the troublesome question of what is to constitute a problem, goal or need, or because they commit the teleological fallacy, all such attempts at definition are in actual fact only heuristics.

A problem, goal or need is a particularly human invention. While humans, nations and cultures may speak of five-year plans to solve their problem, nature (more accurately the complex of heuristics humans personify as nature) does not seem to have seen the need for a 4.5 billion year plan to solve hers. Americans may feel that they have a problem in getting enough energy for their automobiles, but nature sees only an uneven distribution of energy masquerading in a variety of forms, such as kinetic, mass and potential energy, and appears totally uninterested in what this distribution is at any given instant. Similarly, it seems a bit forced to say that a river has a problem in flowing to the sea or that a rock has one in falling. To be sure, arguing that a dog, fish or plant has a problem finding food or shelter seems somewhat more reasonable. On closer inspection, however, even these examples appear suspect.

More accurately, a problem is not a particularly human invention, but a particular human's invention. What passes as a problem for one person may not for another. When water is diverted from rivers to cool large power plants and returned at a higher temperature, some see a problem of thermal pollution, others a blessing of thermal enrichment that enhances the breeding of shrimp and fish. Even such a seemingly uncontroversial problem as stopping war has its detractors, since some people will always profit from the fighting and will not work to end it. The property owner whose land is taken by eminent domain to make way for a highway will see only a problem when someone else's problem is solved. A problem is not a problem until someone thinks it is, and he thinks it is based on the value heuristics in his own sota. This ambiguity in knowing what is to constitute a problem in an absolute sense is the first reason why identification of engineering as problem solution is at best a heuristic.

Weakening the problem-solving aspect of engineering to avoid personal value judgments and replacing it with the more neutral definition of engineering as a goal-directed activity is not much help. In this form, engineering commits the teleological fallacy. Teleology is the study of design in nature. That is, it is the characteristic of nature, or natural processes, of being directed toward a specific end or purpose. In lay terms, it is the notion that the future can somehow affect the present. The question we must now answer is, In engineering, does the future really affect the present, or is the goal-directed aspect of engineering only an illusion based on a lack of understanding of the sophistication of the engineer's sota?

You want a bridge; the engineer will design one for you. You want an automobile; that too is yours. It certainly seems that the engineer works backwards from a goal, that this goal influences the strategy the engineer uses to reach it, that engineering is teleological.

By themselves, these repeated successes at establishing goals and then achieving them might be taken as adding credence to the idea that the future can pull the present toward it. This explanation neglects the complexity of the engineering sota that makes a teleological explanation unnecessary. Two examples will demonstrate this often-forgotten complexity.

Although engineering goals are certainly desirable, they are seldom the most desirable ones that could have been established. Why has an airplane that would go Mach 10, a manned landing on Pluto or a power plant using nuclear fusion not been considered? If engineering were teleological and engineering objectives based exclusively on their desirability, I would be surprised that these more advanced goals were never established and, once established, achieved.

The heuristic needed to explain the engineer's remarkable success rate is that he carefully avoids problems he knows he cannot solve. In effect, the engineer is a good prophet because he makes only self-fulfilling prophecies. In other words, the engineer chooses a project based less on its desirability than its feasibility. The sota of the engineer not only contains heuristics to cause change, but also heuristics to show which changes he can cause. The engineer calls this heuristic the *feasibility study.* In a feasibility study resources are allocated, not with the goal of solving a problem, but with the goal of finding out if a problem is solvable. This goal of determining the feasibility of an idea is achieved whether the final answer is yes or no. Even in a feasibility study, the engineering prophet keeps his reputation intact by considering only goals he knows he can attain. Any engineer alive in 1985 would tell you that neither an aircraft that would go Mach 10, a manned landing on Pluto nor a power plant based on nuclear fusion was feasible at the time. He would never have dared to establish these as engineering goals or, if by chance they were established, to expect them to be achieved.

In actual fact, the matter is far more subtle. An engineer cannot even conceive of goals, much less establish or reach them, if they cannot be expressed in terms of heuristics in his current sota. Before 1905, the engineer could not even suggest the creation of the atomic bomb, because Einstein's heuristic, $E=mc^2$, was unknown. An engineering problem is nothing but a shorthand symbol for a set of current heuristics. This set does not contain any future, presently unknown ones. The engineering prophet is successful because he predicts only what is immanent in the present sota of the engineering profession.

We now have an explanation of the engineer's unexpectedly high success rate that does not commit the teleological fallacy, based on a fuller understanding of the engineering sota. Since the identification of engineering as a goal-directed activity does commit this fallacy, I feel justified in demoting this definition of engineering to a heuristic.

The second example supporting the belief that engineering teleology is unwarranted once the complexity of the engineering sota is taken into account is based on the ability of engineers to solve problems that are originally known to be unsolvable. When the design of an American commercial supersonic airplane was being considered, economic heuristics dictated that it carry at least 200 passengers. The only known material for its outer surface that could withstand the high temperatures generated at supersonic speeds with this payload was titanium. Design continued, although at the time techniques were unavailable for welding this material. Surely this is an example of a goal that was set without knowing essential heuristics that would be needed for its achievement. Actually, it is not. Once again, the paradox results from not appreciating the complexity of the engineering sota.

Recognizing that the sota is a function of time, the engineer does not base the feasibility of a design on the sota that currently exists but on the one he thinks, heuristically, will exist when he needs it. When engineers were considering the supersonic airplane, heuristics were available to predict that within the next ten years techniques would become available to weld titanium. (I might add that these heuristics were good ones, for now, one decade later, titanium is weldable.) The feasibility of the airplane was based on current heuristics for projecting a sota into the future.

Once again, the engineering sota is seen to be far subtler than usually supposed. It contains heuristics for solving problems, heuristics for posing feasible problems and, as we have just seen, heuristics for determining if problems will be feasible in the future. And in all these cases the teleological fallacy is not committed.

Given the ability of an engineering sota to deal satisfactorily with the problem, goal and needs aspects of engineering, as well as the illusion of engineering teleology, I see no alternative but to replace definitions that contain these concepts with the

> ***Heuristic:* Engineering is a problem-solving, goal-directed and needs-fulfillment activity**

or some similar formulation.

Preferred Definition of the Engineering Method

All present efforts to define the engineering method in an absolute way fail. Engineering design is not just a morphology; it is not just applied science; it is not just trial and error; it is not just problem solution, goal attainment or need fulfillment. A more global view of engineering is needed. Establishing this more comprehensive view is the third and last objective of this part of our discussion.

Throughout this discussion, I have repeatedly expressed my preference for the

> ***Heuristic:*** **The engineering method is the use of heuristics to cause the best change in a poorly understood situation within the available resources.**

This does not mean, however, that this choice is any less of a heuristic than those rejected, but only that it is a better one.

This preferred definition of the engineering method may be simplified and put in a more compact form, for its elements either describe an engineering problem situation or are themselves heuristics and hence redundant. *Poorly understood* and the equivalent phrases used in this discussion actually refer to resources, or in this case the lack of resources. Just as with a lack of time or money, a lack of knowledge constrains a problem's solution. This concept may therefore be combined with the word *resource.* Also, the engineer's *best* is not an absolute one, but depends on a complex underpinning of heuristics and may be struck from the definition. *Causing change* and *within the available resources* were used to describe an engineering problem situation and, by implication, one that requires heuristics from the sota of the engineer. As a result, the

> ***Heuristic:*** **The engineering method is the use of engineering heuristics**

is my candidate for a final, compact definition of the engineering method, with the understanding that the phrase *engineering heuristics* is intended to include the specific heuristics outlined above.

This preferred definition of the engineering method is a superior heuristic for five reasons. First, it does not require the engineer at *A* in Figure 1 (page 7) to know the exact final value system that will characterize the future point, *B.* Second, it does not commit the teleological fallacy. Third, the proposed definition is a universal definition of the engineering method; that is, it is always a *good* heuristic in that it always describes what an engineer does. While at times an engineer is using a good heuristic when he establishes a goal or defines a problem based on his current sota and then sets out to solve it, often he is not. A highly hypothetical example will make this point.

Assume that a country needs more energy and has agreed on the goal of building more nuclear power stations. Since energy is needed to train people for reactor design, to mine the materials needed for its construction and to fabricate the final station, it might turn out that to develop nuclear power, the nuclear engineer should head in the exact opposite direction and promote the construction of conventional power stations to ensure that enough energy is available to construct the nuclear plant. And (to preserve the balance in the examples) it might be necessary for a person who is against nuclear power as an ultimate energy source to endorse nuclear power in the short term to ensure that enough energy is available to keep a society stable until his goal of an alternative energy supply could be realized. Unfortunately, establishing a goal is too often taken as a mandate to head straight in its direction. As in the two examples just given, at times this implied mandate is not a very good heuristic, and to achieve our stated goal we may appear to move away from it. Regardless of whether establishing a goal is a good heuristic in a specific instance, whether he establishes a goal or not, the engineer is still using heuristics. In other words, "establish a goal and then try to accomplish it" may on occasion be very bad advice, but "use heuristics" is always a good heuristic to follow.

Fourth, the proposed definition of engineering method includes the conventional ones as special cases. Our definition of the engineering method seeks to find heuristics to help choose the best direction (technically the derivative, of course) at the present moment. Identifying a problem for solution or establishing a goal does just that. They suggest a direction in which to go. They are therefore actually only two of many possible derivative-choosing heuristics.

And finally, the fifth reason why the definition of the engineering method as the use of engineering heuristics is preferable: the derivative, unlike the goal or problem, is not static as we move along the transition from *A* to *B*, but is constantly changing. The lack of information that always plagues an engineering problem suggests that during the transition from *A* to *B* society's value system may well change and a new goal such as B' or B'' may become more desirable. The derivative always represents what is currently thought to be the best policy. For these five reasons, the engineering method is best defined as the use of engineering heuristics.

Figure 11 represents the world of the engineer. He is located at point *A* in time, and his job is to "cause the change" to the "most desirable" final state among all the possible final states represented by the large number of points (such as *B, B′, and B″)* in this figure. Each of these final states is defined by a subset of the engineer's sota evaluated in the present. The proposed definition of the engineering method as the use of engineering heuristics focuses attention on point *A* and the heuristics that define the best direction in which to go, as indicated by the arrow in the figure. The engineer's world is completely defined by the sota $|_{\text{eng.prof.;now.}}$

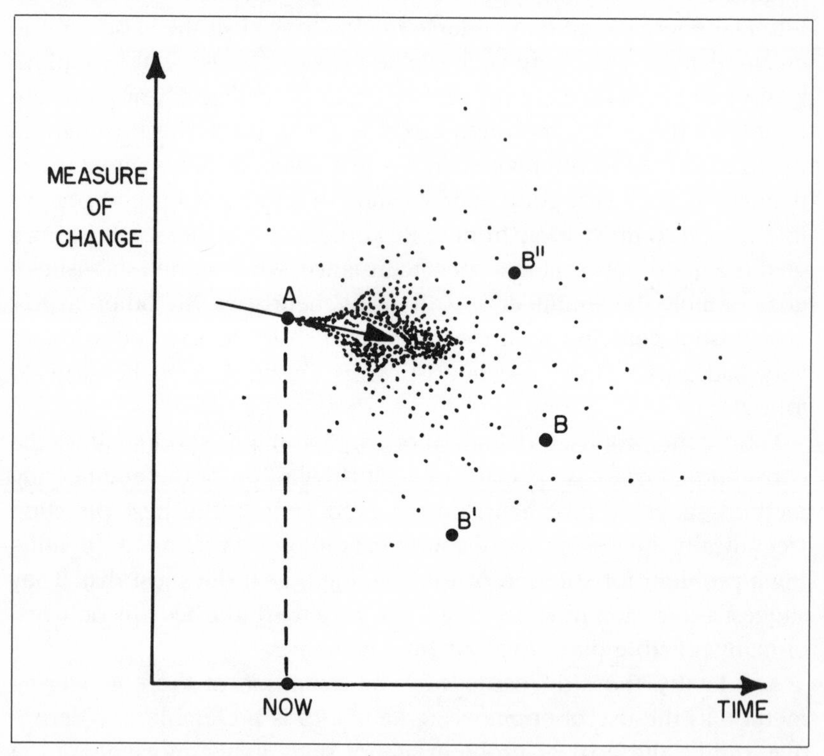

Figure 11

At long last we have concluded our survey of engineering heuristics, including one special heuristic that defines the engineering method. As advertised at the beginning of this section, the list was not intended as a complete feast of all heuristics used by the engineer or even of all the categories into which engineering heuristics may fall. My aim was to encourage you to see the heuristic behind everything the engineer does and to acquire a taste for engineering heuristics beyond the simple ones taken over from the traditional study of problem solution.

From the initial observation made in the introduction that the engineer affects our world incisively, through the discussion of the heuristic and sota, with a pause to review a selected list of engineering heuristics, and so on at last to the final definition of the engineering method, our discussion has revolved around the engineer and what he does. Engineering: the use of engineering heuristics—with this concatenation the engineering profession is no longer obsessed with its artifacts, but becomes concerned with its art. What enormous potential is unleashed as the engineer enters his maturity and uses heuristics (that is, his method) to study the engineering heuristics he will use in the future. The responsibility of each human as engineer becomes clear. It is to develop, learn, discover, create and invent the most effective and beneficial heuristics. The world, as we now know it, vindicates the sota of past engineers; the future world, if we have it, will vindicate this new sota of the present ones.

Acknowledgment

The author would like to thank Deanne Koen for more help in writing, revising and proofreading than any husband has a right to expect. He also owes a debt to the Alfred P. Sloan Foundation, without whose support this monograph would never have been written. In addition, he appreciates the tireless contribution of the ASEE editorial staff and the Society's Publications Committee.

About the Author

Billy Vaughn Koen, professor of mechanical engineering at the University of Texas at Austin, is known in education for introducing the Personalized System of Instruction (PSI) in engineering education and for his research into the theory of engineering design.

Professor Koen received B.A. and B.S. degrees in chemistry and chemical engineering from the University of Texas at Austin; M.S. and D.Sc. degrees in nuclear engineering from the Massachusetts Institute of Technology and a Diplôme d'ingénieur from the Institut National des Sciences et Techniques Nucléaires, France. He is a member of numerous honorary organizations including Tau Beta Pi, Phi Beta Kappa and Sigma Xi and is listed in 20 international and national biographical directories. Dr. Koen has received national, state and local teaching awards including the Chester F. Carlson Award, Minnie S. Piper Professorship and Standard Oil of Indiana Award for Outstanding Teaching. He has been a consultant for the French Atomic Energy Commissión for two years in his specialty of nuclear reactor reliability.

Dr. Koen is currently collecting examples of engineering heuristics of all kinds and solicits such examples from the engineering community.